Shakespeare on the Ecological Surface uses the concept of the 'surface' to examine the relationship between contemporary performance and ecocriticism. Each section looks, in turn, at the 'surfaces' of slick, smoke, sky, steam, soil, slime, snail, silk, skin and stage to build connections between ecocriticism, activism, critical theory, Shakespeare and performance.

While the word 'surface' was never used in Shakespeare's works, Liz Oakley-Brown shows how thinking about Shakespearean surfaces helps readers explore the politics of Elizabethan and Jacobean culture. She also draws surprising parallels with our current political and ecological concerns. The book explores how Shakespeare uses ecological surfaces to help understand other types of surfaces in his plays and poems: characters' public-facing selves; contact zones between characters and the natural world; surfaces upon which words are written; and physical surfaces upon which plays are staged.

This book will be an illuminating read for anyone studying Shakespeare, early modern culture, ecocriticism, performance and activism.

Liz Oakley-Brown is Senior Lecturer in the Department of English Literature and Creative Writing at Lancaster University, UK. Her publications include *The Rituals and Rhetoric of Queenship: Medieval to Early Modern* (co-edited with Louise J. Wilkinson; 2009), *Shakespeare and the Translation of Identity in Early Modern England* (2011) and *Twelfth Night: A Critical Reader* (co-edited with Alison Findlay; 2014).

Spotlight on Shakespeare
Series Editors: John Garrison and Kyle Pivetti

Spotlight on Shakespeare offers a series of concise, lucid books that explore the vital purchase of the modern world on Shakespeare's work. Authors in the series embrace the notion that emergent theories, contemporary events, and movements can help us shed new light on Shakespeare's work and, in turn, his work can help us better make sense of the contemporary world. The aim of each volume is two-fold: to show how Shakespeare speaks to questions in our world and to illuminate his work by looking at it through new forms of human expression. Spotlight on Shakespeare will adopt fresh scholarly trends as contemporary issues emerge, and it will continually prompt its readers to ask, "What can Shakespeare help us see? What can he help us do?"

Spotlight on Shakespeare invites scholars to write non-exhaustive, pithy studies of very focused topics—with the goal of creating books that engage scholars, students, and general readers alike.

Available in this series:

Shakespeare through Islamic Worlds
Ambereen Dadabhoy

Shakespeare on the Ecological Surface
Liz Oakley-Brown

For more information about this series, please visit: www.routledge.com/Spotlight-on-Shakespeare/book-series/SOSHAX

LIZ OAKLEY-BROWN

Shakespeare
on the Ecological Surface

LONDON AND NEW YORK

Designed cover image: © Getty Images

First published 2024
by Routledge
4 Park Square, Milton Park, Abingdon, Oxon OX14 4RN

and by Routledge
605 Third Avenue, New York, NY 10158

Routledge is an imprint of the Taylor & Francis Group, an informa business

© 2024 Liz Oakley-Brown

The right of Liz Oakley-Brown to be identified as author of this work has been asserted in accordance with sections 77 and 78 of the Copyright, Designs and Patents Act 1988.

All rights reserved. No part of this book may be reprinted or reproduced or utilised in any form or by any electronic, mechanical, or other means, now known or hereafter invented, including photocopying and recording, or in any information storage or retrieval system, without permission in writing from the publishers.

Trademark notice: Product or corporate names may be trademarks or registered trademarks, and are used only for identification and explanation without intent to infringe.

British Library Cataloguing-in-Publication Data
A catalogue record for this book is available from the British Library

ISBN: 978-0-367-34591-4 (hbk)
ISBN: 978-0-367-34586-0 (pbk)
ISBN: 978-0-429-32675-2 (ebk)

DOI: 10.4324/9780429326752

Typeset in Joanna
by Newgen Publishing UK

This book is dedicated to all the students in the Department of English Literature and Creative Writing at Lancaster University, UK. Your indefatigable passion for reading, thinking, writing and discussion makes everything I do in Higher Education worthwhile.

Contents

Textual note ix
Acknowledgements x
List of abbreviations xiii

Introduction: surface—now and then 1
Slick: art for what's sake? **One** 25
Smoke: London's burning **Two** 42
Sky: unfirming the firmament **Three** 64
Steam: under pressure **Four** 84
Soil: down to earth **Five** 102
Slime: sensory plays **Six** 121
Snail: finding our place **Seven** 140
Silk: textile production **Eight** 160
Skin: curating complexion **Nine** 186

Stage: disposable globes **Ten** 212

Afterword: surface futures 233

Further reading 236
Index 239

Textual note

All quotations from Shakespeare's plays and poems are from Stephen Greenblatt, Walter Cohen, Jean. E. Howard and Katharine Eisaman Maus (editors), *The Norton Shakespeare*, second edition (New York: W.W. Norton, 2008).

All references to the use of principle words in Shakespeare's plays and poems are from the 'Concordance of Shakespeare's Complete Works', *Open Source Shakespeare: An Experiment in Literary Technology* (George Mason University 2003–2021). www.opensourceshakespeare.org/concordance/

Unless otherwise stated, the spelling of unedited early printed sources has been retained. I have modernised i/j; u/v.

All URLs were accessed on 10 June 2023.

Acknowledgements

I want to begin with an apology. I've been working on Shakespearean Surfaces for such a long time that I know I'm going to overlook some important conversations I've had with so many people, in embodied and virtual forms, over the years. Please know that I'm extremely grateful for every surface-related discussion I've had. Thank you.

I'm indebted to the following friends and colleagues who have helped me shape the ideas for this book. Alison Findlay co-organised the seminar 'Shakespearean Surfaces' with me for the 2007 British Shakespeare Association Conference. In May 2013, Rebecca Coleman co-organised the 'Surfaces in the Making' installation and symposium with me at Lancaster University and co-edited a special section of *Theory, Culture and Society* based on that symposium, 'Visualizing Surfaces, Surfacing Vision', in 2017. In the wake of the talk she gave for 'Surfaces in the Making', I've enjoyed correspondence with Patricia Cahill about her research on affect and early modern stage properties, specifically animal skins. I'm grateful for Kevin Killeen's enthusiasm to co-organise a joint University of York/Lancaster University seminar on 'Scrutinizing Surfaces in Early Modern Thought' (2015) and his expertise in co-editing the follow-up special issue of articles for *The Journal of the Northern Renaissance* (2017). In the early stages of this book, Andrew Gordon invited me to deliver a talk on 'Shakespeare on the Surface' for Aberdeen University's Centre for Early Modern Studies (2019). During the first part of the global Covid-19 pandemic, I was intellectually and emotionally uplifted by everyone who engaged with my hashtag #SurfaceStudies on social media platforms but especially @starcrossed2018 Hester Lees-Jeffries. Much thanks and gratitude to Cat Evans, Lucy Razzall and Emily Rowe for their papers on our jointly organised panel 'Premodern Surface Studies: Paper, Pearl, Patina' for the 2022 British Society for Literature and Science Conference. Lucy Razzall's specific expertise on early modern boxes, paper and cardboard has been an important influence on my work since 2015.

I'm very appreciative of the time Jess Worth, co-founder of BP or not BP?, spent with me to discuss the Reclaim Shakespeare's Company work. Anne Valérie Dulac, Elizabeth Robertson and Kristine Steenbergh very kindly sent me copies of seminar papers or advance copies of their essays. I had the great privilege of learning from Wendy Lennon's important 2022 Black History Month talk—'Skin/Pedagogy'—for Lancaster University. I am beholden to Wendy Lennon and Ruben Espinosa for reading and commenting on drafts of this book's sections, 'Introduction' and 'Skin: Curating Complexion'.

I must thank my colleagues (past and present) in the Faculty of Arts and Social Sciences (FASS) at Lancaster University for their collegial input, which shapes my thinking on a regular basis: Jenn Ashworth, Simon Bainbridge, Sally Bushell, Jo Carruthers, Sarah Casey, Helen Davies, Philip Dickinson, Clare Egan, Kamilla Elliott, Michael Greaney, Lee Hansen, Hilary Hinds, Bethany Jones, Mark Knight, Lindsey Moore, Tony Pinkney, Sharon Ruston, John Schad, Jen Southern, Catherine Spooner, Andrew Tate and Rachel White. I'd like to give extra praise to Michael Greaney for finding space in his packed schedule to read and comment on drafts of this book's sections, 'Introduction' and 'Snail: Knowing Our Place', and to Lee Hansen for invaluable discussions about writing in general.

As I note in the 'Introduction', individual papers by Lucy Razzall and Lawrence Green at the 'Scrutinizing Surfaces in Early Modern Thought' seminar taught me that Shakespeare avoided the use of the word surface. Without them, this book wouldn't exist. This book also wouldn't exist without the professional kindness, generosity and patience of Spotlight on Shakespeare's series editors. I've been so lucky to work with John Garrison and Kyle Pivetti during these challenging times. No words can effectively express my thanks. I'd also like to thank Routledge's Chris Ratcliffe for sending me serene, supportive and pragmatic correspondence. Finally, I'm appreciative of everyone who has listened to me talk *endlessly* about finishing this book and has supported me in numerous ways—not least my partner Rob Douglas.

Abbreviations

ODNB *Oxford Dictionary of National Biography*, www.oxforddnb.com/
OED   *Oxford English Dictionary Online*, www.oed.com/

Introduction

Surface—now and then

Contemporary popular culture is keen to credit Shakespeare, often incorrectly, with the coinage of some English term or another. Instead of celebrating Shakespeare as a point of origin, my book is interested in an omission. *Shakespeare on the Ecological Surface* works out from the curious fact that Shakespearean drama and non-dramatic verse don't use the noun 'surface'. There are words like it—for example, 'superficial' in *Henry VI Part One* (5.7.10) and *Measure for Measure* (3.1.379)—but not 'surface' per se. My book extends awareness of this peculiar gap, an evocative flashpoint which came to light at a conference I co-convened in 2015 on 'Scrutinizing Surfaces in Early Modern Thought'.[1] As a consequence of this linguistic lacuna, three main questions underpin my critical perspective of Shakespeare's plays and poems:

> What are the implications of the surfacing of the word 'surface' itself?
> How does a consideration of Shakespearean surfaces help to explore premodern cultural politics?
> To what extent does thinking about surfaces in Shakespeare's texts and their afterlives put a spotlight on twenty-first century ecological concerns?

Given all the current evidence for climate collapse, and as my book's title suggests, the third question is the most pressing one. From an openly presentist[2] point of view, I tie my exploration of the term and concept of surface to three areas of twenty-first century scholarship: critical medical humanities, environmental studies and social activism. As *Shakespeare on the Ecological Surface* is broadly interested in the naming of surfaces—the moment when 'surface' seems to become a thing itself—my discussion is also aligned with object studies and what some have seen as a developing field of surface studies.[3] Via one of the most canonical set of texts in Anglophone cultures and societies, my book argues that humans live on, with, among and manipulate surfaces and is organised into ten short sections headed by alliterative keyword case studies—Slick, Smoke, Sky, Steam, Soil, Slime, Snail, Silk, Skin and Stage—as a reminder of the relationality between these apparently disparate but, as I try to suggest, necessarily interconnected things.

When Michel Foucault theorised how invisible networks of power disciplined the body, he famously used Jeremy Bentham's nineteenth-century architectural design of the Panopticon to help explain his concept. 'Power', he argues, 'has its principle not so much in a person as in a certain concerted distribution of bodies, *surfaces*, lights, gazes; in an arrangement whose internal mechanisms produce the relation in which individuals are caught up' (my emphasis).[4] In doing so, Foucault suggests that surfaces are important components of social surveillance. And yet, it's tricky to ascertain what constitutes a surface in the first place.

As part of the twenty-first century's general turn to surfaces, surface studies is framed by discussions such as Isla Forsyth, Hayden Lorimer, Peter Merriman and James Robinson's 'What Are Surfaces?'[5] and Mike Anusas and Cristián Simonetti's volume of essays *Surfaces: Transformations of Body, Materials and Earth*.[6] Generally underpinned by William Gibson's and Tim Ingold's respective takes on surfaces as zones of perception and becoming, this compelling collection of essays on topics such as 'air, smoke and fumes in Aymara and Maupache rituals', skin and taxidermy, and knitting[7] show this growing field's dynamism:

> Considering surfaces, knowledge of the world is not that of an optical incision through superficiality to the matters of a fixed depth in waiting but rather that of a responsive sensorial encounter with entanglements of life that are ever moving and growing... Dialoguing with these complementary agendas on sensing and mattering, this volume seeks to overcome dichotomies of modern thought by attending to surfaces not as entities on one side of a division but rather as transformative thresholds which manifest different qualities in the meeting of minds, bodies, materials and earth.[8]

Like Anusas and Simonetti, I'm especially interested in troubling modernity's violent hierarchy of surface and depth, but my approach is by way of Shakespeare. In many ways, *Shakespeare on the Ecological Surface* thinks of the author's plays and poems as 'transformative thresholds' produced in a European epoch, foregrounding a 'responsive sensorial encounter with entanglements of life' that take a back seat with the rise of Cartesian dualism and the Enlightenment's general preference for rationality and order.

One of my greatest intellectual debts in exploring Shakespearean surfaces is to Joseph A. Amato's 2013 book *Surfaces: A History*, a critical/creative transhistorical study which looks at 'our relation to surfaces in order to carry out a historical, philosophical, and anthropological meditation on humans as self-reflecting, self-defining, and self-making creatures'.[9] Along the way, and as a means of considering how 'surfaces, in all their variety, define margins, set down borders, establish grids, and form interfaces' and 'materialize the great juxtaposition between inside and outside', Amato makes two brief but important references to Shakespeare: a quotation from *Hamlet* to exemplify rhetoric's relational capacity to make meaning and a comment on Shakespearean tragedy's ability to show how cultural consensus consolidates meaning and 'turns surfaces into the coin of the realm'.[10] As Amato observes, Shakespeare shows how the slipperiness of meaning in language is held in place by social and political ideologies. My sustained focus on the late sixteenth- and early seventeenth-century writer's work takes up Amato's observations and allows me to take a deeper dive into suggestive connections between critical medical humanities, environmental studies, social activism and object studies as well as how they engage with premodern European outlooks.

The overarching idea of my book, then, is that the surfacing of the noun 'surface' in the vernacular marks a shift in England's relationship with the world. My main point is that the Shakespearean texts' avoidance of the word 'surface' is striking when placed against our own period's fascination with it. (A companion piece to my discussion of sixteenth- and seventeenth-century English writings and cultures is Cynthia Sundberg Wall's brilliant book *The Prose of Things: Transformation of Description in the Eighteenth Century*, which considers how the later period 'demand[ed] to see the surfaces of their worlds'.[11]) I'm not saying that Shakespeare and his contemporaries didn't think about surfaces or try to theorise them. After all, the first published English translation of Euclid's *The Elements of Geometries* (300 BC) by

Henry Billingsley appeared in 1570 and contains remarkable pop-up illustrations to accompany its discussion of three-dimensional geometry.[12] Rather, I'm interested in the fact that the word itself isn't recorded in the English vernacular until Shakespeare's lifetime (the digitised collection Early English Books Online [EEBO] suggests the earliest date is 1581 while the OED says 1594). I'm also gripped by the idea that the term 'surface' starts to circulate in English at roughly the same time that some scientists suppose that the Anthropocene—the geological epoch describing humankind's impact on the earth's climate and environment—began with the so-called Orbis hypothesis and the change in CO_2 circa 1610.[13]

Drawing particular attention to premodern colonisation's propensity for massacre, Philip John Usher, after Simon L. Lewis and Mark A. Maslin, explains the relationship between those deaths and climate change:

> the arrival of Europeans in the New World in 1492 and the subsequent century of slaughter of indigenous populations—whose numbers fell by approximately fifty million—might serve to mark the beginning of the Anthropocene. The [Orbis] hypothesis turns mainly on the fact that the huge number of deaths resulted in a near cessation of farming, a reduction in fire use for land management, the regeneration of over fifty million hectares of forest, savanna, and grassland, and thus in a significant increase in carbon sequestration.[14]

While Usher's book-lengthy study called *Exterranean: Extraction in the Humanist Anthropocene*, as its title suggests, isn't about emission but withdrawal, this account of the Orbis hypothesis starts to emphasise the devastating effects of asymmetrical power relations between human and non-human inhabitants of the earth. Kathryn Yusof's *A Billion Black Anthropocenes or None*, 'a meditation on the politics and poetics of abjection that underpin the becoming of the Anthropocene as a material and durational fact in bodies and environments',[15] shows how those fifteenth-century New World encounters contribute to twenty-first century constructions of race and gender. Usher's work doesn't go as far as Yusof's. Nonetheless, Usher's book mindfully breaks apart the hierarchical binary structure of surface and depth that's been in place since Nicholas Steno's geological concept of stratigraphic time[16] in the late seventeenth century. Such foregoing factors mean that the production of Shakespeare's plays and poems coincide with the rise of exploration, excavation and colonial enterprise enabled by social and economic privilege.

Shakespeare on the Ecological Surface doesn't suggest that the avoidance of the word 'surface' in Shakespeare's plays and poems is a conscious lipogramatic exercise. Figures of speech (simile, metaphor, metonymy, synecdoche and so on) stand in for 'surface' and as such they are culturally contingent and open to interpretation; Shakespearean drama's non-verbal signs (stage directions, set design, props, clothing) also influence that interpretation. Moreover, sixteenth- and early seventeenth-century culture endorses a relational worldview based on the four elements—earth, air, fire, water—which has little need for concepts of surficial division. Such an elemental experience is found in one of Shakespeare's favourite source texts, Ovid's *Metamorphoses*:

the endless world contains four generative
bodies. The two more massive ones of these,
water and earth, sink under gravity,

> The others, rising weightless and unhindered,
> are air and fire, purer than air. Though set
> apart in space. They form out of each other
> and pass into each other. Earth dissolves
> and thins to liquid. Water vaporizes,
> changing to wind and air. Air rarifies,
> losing its weight, and leaps up high as fire.
> Then they repeat this order in reverse:
> Fire condenses, changing into air,
> the water. Liquid hardens into earth.[17]

Shakespeare's works are thus caught up in a shift from a culture that values joined-up elemental thinking to one that so readily thinks with surfaces that the political freight of the word itself has been overlooked. Surfaces help to sort out the multifarious and complex physical issues of being human in the first place, for example the difference between inaccessible and accessible terrains. We can't do without them, or at least the idea of them, but it's worth keeping in mind that the word 'surface' is held in place by visible and invisible systems of power, knowledge and cultural consensus.

Back in the early 2000s, Gabriel Egan revived E.M.W. Tillyard's 1943 discussion in *The Elizabethan World Picture* about the 'the chain of being' (the hierarchical Christian belief system that ordered the world from God down to plants, rocks and minerals) and put it into conversation with James Lovelock and Lyn Margulis' 1970's Gaia principle (the idea that 'the earth acts like a living organism—that life is part of a self-regulating system, manipulating the physical and chemical environment to maintain the planet as a suitable home for life itself'[18]) to make a case for a historically inflected ecocritical approach to Shakespeare.[19] While the idea of planetary thought seems more democratic than say a 'chain of being', thinkers such as Bruno Latour have extended Lovelock and Margulis' Gaia principle as a means of troubling concepts of worlding in the first place:

> he who looks at the Earth as a Globe always sees himself as a God. If the sphere is what one wishes to contemplate passively when one is tired of history, how can one manage to trace the connections of the Earth without depicting a sphere? By a movement that turns back on itself, in the form of a *loop*.[20]

Shakespeare's Ulysses provides some notion of just how difficult it is to shift from the comfort of spherical planetary mindsets to other conceptual modes:

> The heavens themselves, the planets and this centre
> Observe degree, priority and place,
> Infixture, course, proportion, season, form,
> Office and custom, in all line of order.
> ...
>   O, when degree is shaked,
> Which is the ladder to all high designs,
> Then enterprise is sick. How could communities,
> Degrees in schools and brotherhoods in cities,
> Peaceful commerce from dividable shores,
> The primogenity and due of birth,
> Prerogative of age, crowns, sceptres, laurels,
> But by degree stand in authentic place?
> Take but degree away, untune that string,
> And hark what discord follows. Each thing meets
> In mere oppugnancy. The bounded waters

Should lift their bosoms higher than the shores
And make a sop of all this solid globe;
(Troilus and Cressida 1.3.85–113)

Amid a passionate 63-line speech about the Greek army's faults and the dangers of anarchy, Ulysses's image of 'sop' ('a lump of soaked bread')[21] comes close to Latour's reckoning of the earth as less of a solid globe than a 'tissue of globabble'.[22]

But it wasn't language that kick-started my critical interest in Shakespearean surfaces. Shakespeare's comedy *As You Like It* is often singled out for its fusion of non-human and human attributes. Far from being a convenient setting for the play's comic business, *As You Like It*'s Forest of Arden is a crucial character in its own right. Along with Duke Senior and the rest of the exiled court, the audience 'Finds tongues in trees, books in the running brooks, / Sermons in stones, and good in everything' (2.1.16–17). And Andrzej Krauze's poster for Tim Albery's Old Vic production of *As You Like It* (May 1989) brilliantly captures the play's dramatisation of translation and transformation.[23] Blending landscape and skinscape, Krauze's drawing probes *As You Like It*'s approach to the surface: the

> boundary condition that comes into being through the active relation of two or more distinct entities or conditions, the interface [which] may be distinguished from the *surface*. The *sur-face*, as a facing above or upon (*sur-*) a given thing, refers first of all back to the thing it surfaces, rather than to a relation between two or more things.[24]

Krauze's poster emphasises *As You Like It*'s playful interest in bringing things together rather than keeping them apart. In this comedic scenario, 'boundary conditions' are pushed, tested and occasionally collapse.

The same Old Vic season featured two further Shakespearean plays, both interested in bringing humans and the environment into view: *King Lear* (March 1989) and *The Tempest* (October to November 1988). Again, Krauze's posters help us to consider how these adaptations of Shakespeare's generically different works treat surfaces. By contrast with his artwork for Shakespeare's comedy, Krauze's cross-sectional view of *King Lear*'s Dover shows stark demarcations between sea, cliff, vegetation and sky. Such a stratified perspective upholds the tragedy's focus on divisions of class, gender and nationhood, which stem from a sovereign's wish to retire and split his kingdom among his daughters. While two small, slanted lines are suggestive of 4.5's scene with Edgar and Gloucester on the 'chalky bourn' (57) of Dover's cliffs,[25] Krauze's poster depicts the earth's composition as texturised blocks of blue, white and green. At first glance, Krauze's artwork for *The Tempest*, featuring the reassembled fractured skull of a gently weeping ruff-trimmed bird (perhaps emblematic of 'cormorant devouring Time' (*Love's Labour's Lost* 1.1.4)) also bypasses a distinct human form. But there is something undeniably human about Krauze's skeletal image: the ruff works as a synecdoche for Elizabethan England while the small tear nestling in the corner of an eye socket bespeaks humanoid emotion. *The Tempest* reminds us, like Krauze's poster for the play, that 'We are such stuff/ As dreams are made on, and our little life/ Is rounded with a sleep' (4.1.56–8). In the end, humans die. But Krauze's image also recalls plague-masks, ancestral face coverings anticipating the protective layers we were asked to wear at the outbreak of Covid-19: hoped-for protective surfaces between living bodies and corporeal remains.

When I began working on Shakespearean surfaces in 2007,[26] the publication of Emily St. John Mandel's eerily prophetic dystopian novel *Station Eleven* was still seven years away. St. John Mandel's narrative begins with a professional performance of *King Lear* on a winter's night in Toronto with 'The King stood in a pool of blue light, unmoored'[27] as a lethal virus takes hold of North America. Looking back to my first reading of *Station Eleven* in 2014, I couldn't have imagined writing a book on Shakespearean surfaces during a global pandemic. It wasn't until my co-workers and I received swift instruction to leave our offices on Lancaster University's campus in the north west of England on Friday, 21 March 2020 that Covid-19, the by now all-too familiar name for just one strain of coronavirus, took hold of day-to-day life in ways I still can't fully comprehend: hands were washed red-raw; the body's temperature was regularly assessed; anti-bacterial products were spread and sprayed; face masks (fabric or disposable? under or over prescription eyewear?) donned.

The UK's first lockdown (which began on 23 March 2020, three days after we'd been told to vacate our regular places of work) unleashed a sometimes stimulating but often overwhelming opportunity for me to dwell on non-human and human surfaces. The government's daily briefings televised to the UK population between 16 March 2020 and 23 June 2020 habitually opened with the numerical increase in UK death and infection rates over the last 24 hours, sombre statistical reminders of this coronavirus' impact on the domestic population. Graphs compared the rise and fall of the UK's morbidity with the world's. In all efforts to limit the spread of the disease, public and personal surfaces thus became the UK's foci. Questions were asked about the transmission of Covid-19 and surfaces were first in National Health Service (NHS)

England's list of potential fomites.²⁸ Six months after the UK's lockdown began, the *Telegraph*'s short article 'How long does the coronavirus live on surfaces?' sought expert opinion to answer the question. Plastic, stainless steel, copper, cardboard and fabrics were discussed. According to Bharat Pankhania, our 'mindset' should 'be that everything, everyone, everywhere is contaminated. And whatever you handle is a potential risk'.²⁹ Two months later, the UK government's 'Hands, Face, Space' public information campaign extended their original surface-focussed brief to amplify the importance of ventilation as a means of infection control.³⁰ In the winter of 2020, new mutations of the virus appeared. On 21 January 2021, the UK 'had the highest per-capita daily death toll of any other country in the world'.³¹ So far, not so good.

Living with Shakespeare's surfaces during the start of the global pandemic was an extraordinary experience. Moving swiftly from a time that carelessly embraced the freedom of journeying between houses, streets, towns and cities, I became used to Google Maps' timeline telling me I'd travelled the total of a 1-mile radius (for daily exercise, food supplies and emergencies only) in four weeks. Professional and personal social interactions were increasingly enabled by corporate conglomeration and digital privilege via platforms whose names (Zoom, Skype, Crowdcast, MicrosoftTeams) are suggestive of the very kinds of physical experience this phase of the pandemic prohibited: travel, space and assembly. Did it help me to know that 'during the 16th century, a young couple in Stratford-upon-Avon, England, lost two of their children to the bubonic plague. The pair barricaded themselves inside to protect their 3-month-old son—William Shakespeare. The legendary playwright's life was shaped by the plague'?³² It's hard to say. Nonetheless, I agree with James Shapiro that on the few occasions Shakespeare 'does mention plague, mostly in his tragedies [for instance in *Romeo and Juliet*, *King Lear*, *Macbeth*, *Antony and Cleopatra*, *Coriolanus* and *Timon of Athens*] it hits with incredible force'[33] and I often wonder if this family of sixteenth-century Stratfordians ever felt as acutely aware of surfaces as I am now. In any case, Shakespeare's plays and poems avoid using the specific word 'surface' in its singular or plural forms and so they turn to figures and dramaturgical devices to represent humankind's engagement with the world in ways that sometimes seem far more progressive than my own.

Though separated by nearly half a millennium of social and cultural differences, sixteenth-century plagues and twenty-first century pandemics expose humankind's physical vulnerabilities and the need for protection. It's perhaps no coincidence that one of the earliest books about epidemiology, Girolamo Fracastoro's *De Contagione et Contagiosis Morbis et eorum curatione* [On Contagion, Contagious Diseases and Treatment] published in Venice in 1546, put forward a materialist theory which seemed to comment for the first time on 'fomes' or 'fomites', inanimate objects which conveyed 'seedlets of contagion'.[34] With a partially atomistic perspective, Fracastoro's early germ theory is largely at odds with prevailing humoral sensibilities which, after Galen and Hippocrates, saw human bodies connected to their environment by means of a quartet of liquids or humors (blood, phlegm, black bile and yellow bile) controlling health, emotions and appearance. According to humoral theory, the key to good health involved keeping a balance of each humor through diet and, where necessary, by purgation (vomiting, defecation, bloodletting). By contrast with Fracastoro's ideas, humoralism depends on the body's quite literally fluid relationship with its surroundings. In this premodern culture of confluence, non-human agents and human corporealities are enmeshed: they form assemblages not discrete sites of difference. And in these senses, premodern European culture can be seen as an ecological one before the letter.[35]

Most of my days in the first few months of the pandemic began with reading a small section from Lia Leendeertz's *The Almanac: A Seasonal Guide to 2020* followed by some short, related exercise such as noting the usually neglected shifts in the leaf count on the tree (bronze maple?) outside the kitchen window of my first-floor gardenless flat. I remember circling *The Almanac*'s parenthetical comment for its introduction to 'May', which told me that '(Despite the title, Shakespeare's *A Midsummer Night's Dream*—which has put about the idea that midsummer is when the fairies and humans intermingle—is actually set on May Day Eve)',[36] a small but timely reminder that while Shakespeare might continue to speak to and/or indeed influence twenty-first century thinking, Shakespearean ideas are mediated by those who interpret them: the words of the plays and the poems don't speak for themselves. And part of Shakespeare's enduring appeal is that the texts leave room for interpretation, adaptation and discussion. As an educator working in literary studies, words and how they're used in past and present cultures will always be my first point of reference. For me, surface's silence in Shakespeare's works is a silence worth further consideration.

But how?

At the time of writing this introduction, the UK is in the second month of its third lockdown. I'm still working from home and have spent over a year geographically distanced from my loved ones. Inevitably, this book contains a rather different discussion than the one I originally set out to write, not least in terms of my approach. With physical and virtual travel to local, national and international archives severely restricted—so much is dependent on whether my institution has a subscription to this or that platform—I've become increasingly aware of social media's significance for sharing all kinds of Shakespearean-related ideas, intellectual encouragement and political perspectives. Though Twitter (renamed X in July 2023) can be viewed as a site of procrastination (at best), the accounts I follow on my carefully curated timeline (the use of the 'block' and 'mute' functions are essential) provide vital world views about equality, diversity and inclusion.

Academics like me have been trained to be objective, to be mindful of subjectivity and, above all, to avoid self-indulgence. But during this pandemic, I've also learned that my 'positionality'[37]—a term coined by sociologists in the 1960s—is a crucial component of my critical points of view and the 'production of knowledge' in general.[38] Trying to think and write in this strange space of Covid-enforced physical isolation has emphasised how my professional and personal perspectives are shaped by my privileged position as a middle-class white woman currently in full-time employment in a UK higher education institution. To be sure, even my isolation is a privilege. While I make no claim for *Shakespeare on the Ecological Surface* as autoethnography or memoir—genres which describe very particular modes of embodied encounters—my critical stance in this book is like no other I've taken before: there's a lot of me in this discussion. By adopting what Jessica Soedirgo and Aarie Glas have termed 'active reflexivity', a 'disposition toward both ongoing reflection about our own social location and ongoing reflection on our assumptions regarding others' perceptions',[39] I'm aiming to work with a critical self-awareness that, inevitably, is open-ended.

Despite the three-month advertising partnership between the UK government and the newspaper industry, which began on 17 April 2020 and proposed that the nation was 'All in, all together',[40] 2020's pandemic highlighted—and continues to highlight—the inequalities between those with and without economic and ideological privileges. As the Institute for Fiscal Studies' Paul Johnson put it on 27 April 2020:

> It's tempting to believe that we are all in this crisis very much together. It doesn't matter how rich you are, there is no vaccine you can buy, no cure that is available to you. That's not to say that when we have more data we won't find that death rates have varied by social background. People differ in their ability to socially distance, in where they live, in underlying health conditions. There is certainly evidence of some ethnic groups being more affected than others, perhaps in part because they are more likely to live in cities such as Birmingham and London where there are more cases.
>
> All of this will become apparent over time, but whether or not mortality rates differ by social background, one thing is for sure. We are not all in this together when it comes to the social and economic consequences of the virus and our response to it.[41]

Echoing Shakespeare's own travel route between Stratford-upon-Avon and London, the Institute for Fiscal Studies' report invokes the narrow geographical span of Birmingham (just 63 kilometres from Strafford-upon-Avon) and England's capital city. If Covid-19 has drawn attention to social inequalities,

Johnson's report shows how the pandemic simultaneously foregrounds our planet's precarity.

While the lasting effects of Covid-19 on the earth's short- and long-term well-being are unknown, at the end of August 2020 British media headlines suggested that the month's 'abnormal wildfires'[42] in California and Greenland's 'unprecedented ice loss'[43] were just two clear examples of humankind's harmful impact upon the planet. For some, Covid-19 and climate collapse are very clearly connected. The World Wildlife Fund paper, published in June 2020, warned that 'environmental factors' are 'driving the emergence of zoonotic diseases'.[44] In the same month, the UK's independent statutory body the Committee on Climate Change recommended that 'Ministers seize the opportunity to turn the Covid 19 crisis into a defining moment in the fight against climate change'.[45] The idea that Covid-19 provides a catalyst for environmental action and cultural innovation has been taken up by other public-facing organisations. The press release in August 2020 for the UK's annual Royal Institution's Christmas lectures, for instance, announced that the year's celebrated series of talks (to be delivered by oceanographer Dr Helen Czerski, geologist Professor Chris Jackson and environmental scientist Dr Tara Shine) was called 'Planet Earth: A Users Guide' and designed to 'take audiences on a deep dive into our planet's workings, from rock formation and Earth's ancient climate, to the fundamental role of the oceans and the makeup of the air we breathe'.[46] Tara Shine explained that the Royal Institution's lectures aimed to

> paint a picture of how our planetary system works and where we are as one species within that...Covid is a restart button...To be resilient to the next pandemic we have to build

> some of the same core skills and capabilities that we need to be resilient to climate change.⁴⁷

Helen Czerski added that

> The scientists and the policymakers who are concerned about climate change are extremely sympathetic to the massive suffering a huge number of people have had through this pandemic...The entire point is that this is what societal change looks like when something changes.⁴⁸

Like the UK government's slogan 'we're all in this together', the Royal Institution's speakers' talk of humans as 'one species', acknowledgement of 'massive suffering' and 'societal change' seems inclusive. Only time will tell just how unified earth's community really is.

Alongside oceanographers', geologists' and environmental scientists' views, communities who work with creative works like (but not exclusively) Shakespeare's also have an important role in drawing attention to personal, public and planetary well-being. The Jacobean tragedy *Macbeth* makes a good starting point for thinking about Shakespearean surfaces now and then. Famously, *Macbeth*'s central protagonist is a man who believes he is destined to become sovereign. Unlike the onstage community, the theatre audience knows that Macbeth murdered his King and that innocent lower-class men will take the blame. By comparison with Macduff's loss of articulacy at a 'horror' which 'Tongue nor heart cannot conceive nor name' (2.3.59–60), Macbeth's description of Duncan's corpse relies on a precise kind of scrutiny:

> Here lay Duncan,
> His silver skin laced with his golden blood,

> And his gashed stabs looked like a breach in nature
> For ruin's wasteful entrance; there the murderers,
> Steeped in the colours of their trade, their daggers
> Unmannerly breeched with gore. (2.3.108–113)

If comedy is about toying with surfaces (as my brief discussion of *As You Like It* above suggests), tragedy depicts violent ruptures in human and non-human bodies. Shakespeare's contemporary Philip Sidney put it like this: 'Tragedy... openeth the greatest wounds, and showeth forth the ulcers that are covered with tissue; ... [and] teacheth the uncertainty of this world, and upon how weak foundations gilden roofs are built'.[49] While there is something preternatural in Macbeth's glistening chromatic illustration of regicide, his lines are full of privilege: Duncan's 'silver skin' shines bright with economic and European whiteness. Macbeth's forensic study of the fatal wounds, the 'gashed stabs' which 'looked like a breach in nature', points to the wider implications of the sovereign's death. This is not just the murder of an individual but of God's representative on earth. Macbeth's use of the homophones 'breach' and 'breeched' connects Duncan's wounds to the dagger used to kill him: the object so theatrically conjured by Macbeth in 2.1's solitary speech prefacing the murder itself. All in all, Macbeth delivers a persuasive report of the King's demise. His account of 'the murderers, / Steeped in the colours of their trade' is a shimmering representation of regicide and—as the audience knows—a crafty manipulation of surfaces to create a lie. At the other end of the play, Macbeth's own belief in the shape-shifting properties of the earth's surface itself—that Birnam Wood rather than Malcolm's disguised soldiers comes to Dunsinane in 5.5—ultimately leads to his death.

Timothy Morton's *The Ecological Thought* says that

> The ecological crisis we face is so obvious that it becomes easy—for some, strangely or frighteningly easy—to join the dots and see that everything is interconnected. This is *the ecological thought*. And the more we consider it, the more our world opens up.[50]

My book proposes that Shakespeare's omission of a word which is now so quotidian helps to 'join the dots' and reminds Anglophone cultures (if they still need it) that humankind is part of the earth and not separate from it; that we must do better at working together. Rather than making a definitive pronouncement on its titular topic, my hope is that *Shakespeare on the Ecological Surface* offers a provocation for further conversations about epidemics, environments and political encounters now, then and those to come.

NOTES

1 The conference was organised by Kevin Killeen (York University) and Liz Oakley-Brown and held at Lancaster University, UK, 8–9 May 2015. I gratefully thank and acknowledge two talks in particular: Lawrence Green, "This loam, this rough-cast and this stone": Walls both 'wicked' and 'courteous' in Shakespeare's plays' and Lucy Razzall, 'The other syde of the lefe': Titles, Pages, Surfaces and the Early Modern Material Text'. These excellent presentations considered Shakespeare's lack of engagement with the term 'surface'. A selection of revised papers from the conference was published as a special issue of *The Journal of the Northern Renaissance*, 8 (2017) www.northernrenaissance.org/issues/issue-8-2017/.

2 Hugh Grady and Terence Hawkes (eds), *Presentist Shakespeares* (London: Routledge, 2007).

3 Rebecca Coleman and Liz Oakley-Brown, 'Visualizing surfaces, surfacing vision: introduction', *Theory, Culture & Society*, 34.7–8 (2017): 5–27, p. 9. Some key publications in this area are Joseph A. Amato, *Surfaces: A History* (Berkeley: University of California Press, 2013); Mike Anusas

and Cristián Simonetti (eds), *Surfaces: Transformations of Body, Materials and Earth* (London: Routledge, 2020); Yeseung Lee (ed), *Surface and Apparition: The Immateriality of Modern Surface* (London: Bloomsbury, 2021); Avrum Stroll, *Surfaces* (Minneapolis: University of Minnesota Press, 1984).

4 Michel Foucault, *Discipline and Punish: The Birth of the Prison*, translated by Alan Sheridan (London: Penguin, 1977), p. 202.

5 Isla Forsyth, Hayden Lorimer, Peter Merriman and James Robinson, 'What are surfaces?, *Environment and Planning A: Economy and Space*, 45. 5 (2013): 1013–20.

6 Anusas and Simonetti. 'Introduction', in *Surfaces: Transformations of Body, Materials and Earth*, pp. 1–13.

7 Juan Carlos Skews and Debbie Guerra, 'Air, Smoke and Fumes in Aymara and Maupache Rituals'; Petra Tjitske Lashove, 'Re-Animating Skin: Probing the Surface in Taxidermic Practice'; Lydia Maria Arantes, 'On knitted surfaces-in-the-making', in *Surfaces: Transformations of Body, Materials and Earth*, pp. 29–45, 63–79, 152–66.

8 Anusas and Simonetti. 'Introduction', in *Surfaces: Transformations of Body, Materials and Earth*, pp. 2–3.

9 Amato, *Surfaces: A History*, p. 17. I am also particularly indebted to the following studies for their general approaches to their objects/subjects: Jo Carruthers and Nour Dakkak (eds), *Sandscapes: Writing the British Seaside* (Basel, Switzerland: Palgrave Macmillan, 2020); Jeffrey Jerome Cohen, *Stone: An Ecology of the Inhuman* (Minneapolis: University of Minnesota Press, 2015); Rebecca Coleman, *Glitterworlds: The Future Politics of a Ubiquitous Thing* (London: Goldsmiths Press, 2020); John Garrison, *Glass* (New York: Bloomsbury, 2015); Mark Kurlansky, *Paper: Paging Through History* (New York: W. W. Norton, 2016).

10 Amato, *Surfaces: A History*, p. 30, p. 35.

11 Cynthia Sundberg Wall, *The Prose of Things: Transformation of Description in the Eighteenth Century* (Chicago, IL: University of Chicago Press, 2006), p. 6.

12 This short clip posted by The Smithsonian Libraries and Archives on 29 August 2022 provides a glimpse of the 1570's book: https://twitter.com/i/status/1564218185913696256.

13 Eric C. Ellis, *Anthropocene: A Very Short Introduction* (Oxford: Oxford University Press, 2018), p. 96.

14 Philip John Usher, *Exterranean: Extraction in the Humanist Anthropocene* (New York: Fordham University Press, 2019), p. 12.
15 Kathryn Yusoff, *A Billion Black Anthropocenes or None* (Minneapolis: University of Minnesota Press, 2018), p. 103.
16 Ellis, *Anthropocene*, p. 35.
17 Ovid, *Metamorphoses*, translated by Stephanie McCarter (London: Penguin, 2022), 15.256–68.
18 Ian Enting, 'Gaia theory: is it science yet?', *The Conversation*, 12 February 2012. https://theconversation.com/gaia-theory-is-it-science-yet-4901 para 1.
19 Gabriel Egan, 'Shakespeare and ecocriticism: The unexpected return of the Elizabethan world picture', *Literature Compass*, 1 (2003): 1–13.
20 Bruno Latour, *Facing Gaia: Eight Lectures On The New Climate Regime*, translated by Catherine Porter (Cambridge: Polity, 2017), pp. 136–7.
21 *The Norton Shakespeare*, marginal note.
22 Latour, *Facing Gaia: Eight Lectures On The New Climate Regime*, p. 130.
23 Copies of Andrzej Krauze's posters for the Old Vic's season of Shakespeare are archived in the *Victoria and Albert Museum*, London. https://collections.vam.ac.uk/item/O1166629/poster-krauze-andrzej/.
24 Branden Hookway, *Interface* (Cambridge, MA: The MIT Press, 2014), p. 12. Cited in Liz Oakley-Brown, 'Translating Surfaces: Shakespeare's *As You Like It*, 1599–1989', *The Journal of the Northern Renaissance*, (2017). www.northernrenaissance.org/translating-surfaces-shakespeares-as-you-like-it-1599-1989/#n1 para 1.
25 I'm grateful to Simon Bainbridge for pointing out the significance of these marks.
26 I'd like to thank Alison Findlay for co-convening the 2007 British Shakespeare Association seminar with me on 'Shakespearean Surfaces' and allowing me to join the local organising committee for the 2012 British Shakespeare Conference *Shakespeare Inside-Out: Depth/Surface/Meaning* held at Lancaster University.
27 Emily St. John Mandel, *Station Eleven* (London: Picador, 2014), p. 3.
28 'Definition of fomite', *NHS England*, 29 April 2021. www.england.nhs.uk/coronavirus/primary-care/about-covid-19/virus-transmission/.
29 Eleanor Steafel, 'How long does the coronavirus live on surfaces?', *The Telegraph*, 16 September 2020. www.telegraph.co.uk/health-fitness/body/does-coronavirus-live-different-surfaces/ para 4.

30 'New film shows importance of ventilation to reduce spread of COVID-19', *Department of Health and Social Care*, 18 November 2020. www.gov.uk/government/news/new-film-shows-importance-of-ventilation-to-reduce-spread-of-covid-19.

31 Alexander Smith, 'Britain's Covid daily death toll is one of the worst in the world. What went wrong?', *NBC News*, 22 January 2021. www.nbcnews.com/news/world/britain-s-covid-daily-death-toll-one-worst-world-what-n1255261 para 3.

32 Robin Young and Allison Hagan, '"He didn't flee": Shakespeare and the Plague', *WBUR*, 6 April 2020. www.wbur.org/hereandnow/2020/04/06/shakespeare-plague-coronavirus paras 1–2.

33 Young and Hagan, '"He didn't flee": Shakespeare and the Plague', para 9.

34 Vivian Nutton, 'The reception of Fracastoro's theory of contagion', *Osiris*, 6 (1990): 196–234, p. 200.

35 According to the OED, the term 'ecological' in a biological context, meaning 'Of, relating to, or involving the interrelationships between living organisms and their environment' came into use in the late 1800s. As an adjective describing environmental issues, 'ecological' was first used in the 1960s.

36 Lia Leendeertz, *The Almanac: A Seasonal Guide to 2020* (London: Mitchell Beazley, 2020), p. 111.

37 'The occupation or adoption of a particular position in relation to others, usually with reference to issues of culture, ethnicity, or gender'. OED 2.

38 For an insightful discussion of positionality in practice, see Jessica Soedirgo and Aarie Glas, 'Toward active reflexivity: positionality and practice in the production of knowledge', *PS: Political Science and Politics*, 53.3 (2020): 527–31.

39 Jessica Soedirgo and Aarie Glas, 'Toward active reflexivity', p. 527.

40 Mariella Brown, '"All in, all together": UK government partners with newspaper industry on Covid-19 ad campaign', *Society of Editors*, 17 April 2020. www.societyofeditors.org/soe_news/all-in-all-together-uk-government-partners-with-newspaper-industry-on-covid-19-ad-campaign/.

41 Paul Johnson, *The Institute for Fiscal Studies*, 27 April 2020 https://ifs.org.uk/articles/we-may-be-together-doesnt-mean-we-are-equally paras 1–3.
42 Alastair Gee and Dani Anguiano, 'The climate crisis has already arrived. Just look to California's abnormal wildfires', *The Guardian*, 21 August 2021. www.theguardian.com/commentisfree/2020/aug/21/the-climate-crisis-has-already-arrived-just-look-to-californias-abnormal-wildfires.
43 Matt McGrath, 'Climate change: "Unprecedented" ice loss as Greenland breaks record', *BBC News*, 20 August 2020. www.bbc.co.uk/news/science-environment-53849695.
44 World Wildlife Fund, 'COVID 19: urgent call to protect people and nature', 17 June 2020. wwf.panda.org/wwf_news/press_releases/?364416%2FConditions-are-rife-for-next-pandemic-unless-urgent-action-is-taken-WWF-warns para 1.
45 'Reducing UK emissions: 2020 Progress Report to Parliament', *Climate Change Committee*, 25 June 2020.www.theccc.org.uk/publication/reducing-uk-emissions-2020-progress-report-to-parliament/#outline.
46 Nicola Davis, 'Covid is "restart button" for climate action, Royal Institution Christmas lecturers say', *The Guardian*, 22 August 2020. www.theguardian.com/science/2020/aug/22/covid-climate-change-royal-institution-christmas-lectures-planet-earth-users-guide-pandemic para 2.
47 Davis, 'Covid is "restart button" for climate action, Royal Institution Christmas lecturers say', para 4, para 14.
48 Davis, 'Covid is "restart button" for climate action, Royal Institution Christmas lecturers say', para 16, para 17.
49 Sir Philip Sidney, *A Defence of Poetry* [1582], edited by J. A. Van Dorsten (Oxford: Oxford University Press, 1966), p. 45.
50 Timothy Morton, *The Ecological Thought* (Cambridge, MA: Harvard University Press, 2010), p. 1.

Slick

Art for what's sake?

One

Twentieth-century space travel has allowed humans to see earth in a way they could only previously imagine. A photograph taken by *Apollo 17*'s astronauts on 7 December 1972, known as 'The Blue Marble', and the image captured by *Voyager 1* on 14 February 1990, called 'Pale Blue Dot', show how our planet is defined by a striking colour enabled by the vast amount of liquid water on its outermost layer. By the time deep-sea oil spills are visible on earth's global ocean as glistening slicks, environmental damage has been done and not just in terms of the immediate tragedy.

One of the most devastating events of this kind was the one caused by the *Deepwater Horizon* rig's explosion in April 2010. During the next three months, '300 Olympic-sized swimming pools of oil [leaked] into the Gulf [of Mexico's] waters, making it the biggest oil spill in United States history'.[1] Research published by *Environmental Toxicology and Chemistry* on 18 February 2021 explains that 'the effects of the Deepwater Horizon spill were larger and longer-lasting than previously estimated—capable of disrupting entire food webs and even affecting animals on land'.[2] The fossil-fuel industry's adverse effects on our planetary health are often so insidious that they are easily overlooked. In the words of Heidi Scott, 'Literary critics call our modern state an "oil ontology", which is to say that is has become so omnipresent, we can't

DOI: 10.4324/9780429326752-2

even see oil because we look through oil glasses, or oil eyes'.[3] Preventative rather than reactive solutions are needed before oozing seams of petroleum force humankind to confront the environmental disasters made manifest by oily pools on the surface of the seas. The problem is that we are all too reliant on 'black gold', the informal idiom for petroleum, which bespeaks both its value and allure in a capitalist economy. This section starts with a small but hopeful moment and looks at how Shakespeare might play a part in helping humankind avoid climate catastrophe.

On 2 October 2019 (and with two years of the agreement remaining), the Royal Shakespeare Company announced that their planned eight-year partnership with the London-based multinational oil and gas company BP plc (known as British Petroleum until 2000) was terminating at the end of the year. The work of Shakespeareans might seem far removed from environmental activism. However, the unceasing efforts of the Reclaim Shakespeare Company (hereafter RSC)—a group of actor-vists formed in 2012 'aghast that [their] beloved Bard's works and memory had been purloined by BP in a case of greenwash most foul'[4]—show what focused, informed and subject-specific campaigns can achieve. Part of the larger group of BP *or not BP?* players whose main aim is to draw public attention to the oil industry sponsorship of large-scale creative practices in the UK, the RSC's work help to show how twenty-first century Western culture's appropriation of Shakespearean texts are enmeshed in climate change history. Just one example to place next to the Royal Shakespeare Company's oil-infused epoch is to understand how Washington's Folger Library is bound to the American Petroleum Institute by way of Henry Folger's full-time 'occupation as an oil company executive'.[5] Once a few links between Shakespeare, theatre performance, archives and the oil industry are shown, our individual complicity in supporting that connection by going to see an oil-company-sponsored play or consulting an archive underwritten by 'black gold' becomes much clearer.

BP or not BP?'s campaign had a very clear sense of how to engage the Royal Shakespeare Company's audience and show them via the medium they enjoyed how the arts and oil were inextricably connected.[6] When the Royal Shakespeare Company took up BP's sponsorship in 2011, that corporate support helped subsidise a ticket scheme for '16 to 25-year olds [which] had allowed tens of thousands of young people to see world-class performances'.[7] On 23 April 2012 (the date commemorating Shakespeare's 448th birthday and just past the two-year anniversary of the *Deepwater Horizon* disaster), the RSC's first two-handed, two-minute, 37-line, self-defined 'guerrilla'[8] performance took place immediately before the evening's production of the Jacobean sea-bound play *The Tempest*:

Performer One:
Ladies and Gentlemen, there will now be a two-minute performance by the Reclaim Shakespeare Company.

Performer Two:
What country, friends, is this? [*raises programme*]
Where the words of our most prized poet
Can be bought to beautify a patron
So unnatural as British Petroleum?

Strange association! [*Performer One unveils image of BP's Deepwater Horizon drilling disaster*]
They, who have incensed the seas and shores
From a dark deepwater horizon

Who have unleashed most foul destruction *[Performer One
 unveils image of tar sands]*
Upon far Canada's aged forests,
Clawing out the lungs of our sickening earth

Who even now would bespoil the high, white Arctic *[Performer
One unveils image of untarnished arctic]*
In desperate search of more black gold
To make them ever richer. These savage villains![9]

Taking the stage just before a performance of *The Tempest*, the RSC's creative protest opens with the familiar quotation from *Twelfth Night*: another play in the Royal Shakespeare Company's 2012 season Shakespeare's Shipwreck Trilogy. In doing so, the RSC drew particular attention to the significance of 'wetscapes'[10] for late sixteenth- and early seventeenth-century audiences while simultaneously addressing the ongoing (and as it turns out enduring) *Deepwater Horizon* environmental crisis. The first half of the RSC's brief intervention also foregrounds the global reach of BP's quest for oil: deep tar extraction in Canada; seascape and landscape drilling in the Atlantic Ocean and Alaskan Arctic.

The second half of the RSC's production turns to BP's corporate branding and the political significance of BP's insignia is deftly unpacked by way of Shakespearean dramaturgical conventions:

And yet—

They wear a painted face of bright green leaves,
Mask themselves with sunshine.
And with fine deceitful words
They steal into our theatres, and our minds.

> They would have us sleep.
> But this great globe of ours is such stuff as dreams are made on.
> Most delicate, wondrous, to be nurtured
> For our children and theirs beyond.
>
> Let not BP turn these dreams to nightmares.
>
> Fuelling the Future? Thou liest malignant thing! [*holding up programme, looking at back page*]

Inspired by the sun god Helios with whom the BP logo shares its name, the image's interlocking colours (dark green, light green, yellow and white) represent 'heat, light and nature'[11] while its 18 points correspond to the harmonious principles of 'the ancient Chinese art of Feng Shui…because it is a multiple of three'.[12] In their description of the logo as 'a painted face of bright green leaves', the RSC illuminates the hypocrisy of BP's relationship to positive environmental signs and symbols. Invoking that moment from *The Tempest* in 4.1 when Prospero makes the striking analogy between theatrical performance and being in the world, the RSC likewise captures our planet's extraordinary composition and variety: its 'stuff'. The phrase 'fuelling the future' recalls BP's slogan as the official oil and gas partner for the 2012 London Olympics, an initiative which included an onsite audio-visual installation which aimed to explain the company's commitment to sustainability.[13]

Shakespeare's plays have always been bound with commerce of some kind. What else are the professional acting companies of the playwright's own time about if they're not concerned with securing and making money? Yet one period's economic market is not the same as another. Susan Bennett's

2016 essay 'Sponsoring Shakespeare' extends Kate McLuskie and Kate Rumbold's discussion of how 'Shakespeare's value is constructed and conferred in commercial settings' by focussing on 'the notion of value in practices of corporate sponsorship'.[14] Bennett thus examines Shakespeare's role in London's 2012 Cultural Olympiad and along the way appraises the RSC's push-back at BP's financial backing. On stage in Stratford, their use of the Royal Shakespeare Company's theatre programme as both prop and addressee allows the RSC to show the audience the kind of short-span leisure experience they have purchased and its long-term environmental cost. With the Macbethian flourish of 'Out damned logo!', Performer Two 'rips out' BP's Helios 'from the programme'. The short vignette concludes with Performer One asking the audience to 'join' them and 'help' them to 'free the arts from BP'. Finally, the excised logos are collected in buckets at the end of *The Tempest*, permitting the actor-vists to count how many audience members share their views while sustaining their environmentalist agenda through this everyday image of recycling. The RSC's performance was so effective that Clare Brennan foregrounds their activism at the beginning of the *Guardian*'s review of the Royal Shakespeare Company's production of *Twelfth Night* on 25 April 2012:

> Before the house lights went down, two men climbed on to the stage and began to sing. Their voices wavered, harmony faltered; they exited auditorium-wards. An usher took their place to explain that this was not part of the performance but an unwelcome protest against BP's sponsorship of the RSC. Embarrassing as it was, the intervention held a certain charm. For all their nervous off-keyness, the singers' performance was touchingly innocent, and the usher's

speech was engagingly forthright—all fitting qualities for
Shakespeare's *Twelfth Night*.[15]

Clearly, the RSC's prefatory piece described as 'touchingly innocent' hit just the right performance mode, at least for this theatre critic. And can it really be a coincidence that the review's accompanying performance still showing Ferdinand (Solomon Israel) and Miranda (Emily Taaffe) as 'Prospero discovers' them 'playing at chess' (5.1.173sd) features a dress of BP green? Once you're alert to BP's influence on the Royal Shakespeare Company's 2012 season, it's difficult to imagine that its costume design is merely happenstance.

According to Caroline Spurgeon, 'Shakespeare's sea images are chiefly of storms and shipwrecks'.[16] In these scenarios, the weather is often viewed as the destructive effect on travel and trade. *The Merchant of Venice*, for example, works out from the possibility that Antonio's enigmatic sadness stems from the loss of one of his ships at sea. As this play's narrative unfolds, the cultural politics of human and economic loss are placed side-by-side, thus enabling the important scrutiny of social inequalities. Nevertheless, overt maritime tragedy isn't *The Merchant of Venice*'s focus. And in what is perhaps Shakespeare's most watery work, *Pericles* emphasises the sea's symbolic and material importance. Steve Mentz explains:

> the sea has not always meant the same things to all people…The early modern sea was not (yet) the sublime theater of crisis and catastrophe that it became in Byron's poems and Turner's paintings. Rather, … early modern literary culture responded to the transoceanic turn of European culture by exploiting the sea's symbolic opposition to and inversion of the orderly world of land. For many early

modern writers, the land is orderly and human; the sea chaotic and divine... As the fishermen in *Pericles* know, however, the ocean is also a space of abundance and recovery. When the waves cast up Pericles, the fishermen insist that their maritime labor provides everything he needs: 'flesh for holidays, fish for fasting days, and moreo'er puddings and flapjacks' [5.115–16]. They even find armor in which Pericles can joust. The combination of hostility and fertility that the fishermen describe captures the sea's role as a metaphor for the contingencies of mortal life from classical and Biblical culture through (and beyond) the early modern period.[17]

For *Pericles*, as with all these Shakespearean oceans that represent chaos and divinity, the sea's ability to provide sites of 'abundance and recovery' for humankind's consumption is not at risk. Likewise, in *The Comedy of Errors*, *Twelfth Night* and *The Tempest*, the earth's waters remain unscathed. In the twenty-first century, we can look back and appraise the cost of early modern flotsam and jetsam—a poetic legal phrase for debris discarded at sea—interacting with maritime environments.

While Shakespeare's plays and poems use oil imagery to express ideas of human dissipation (Edmund Mortimer in *Henry VI Part One* speaks of 'eyes, like lamps whose wasting oil is spent/ wax dims' (2.5.8–9); in *Venus and Adonis*, the goddess tries to persuade the indifferent youth with the lines 'Be prodigal. The lamp that burns by night/ Dries up his oil to lend the world his light' (755–6)), their own period of production is some 300 years before the modern oil industry's explorative, extractive, refinable and marketing processes gets underway. One of the RSC's great strengths is showing audiences how the Royal Shakespeare Company enables British Petroleum's marketing strategies. On 27 April 2017, just before the evening performance of *Antony and Cleopatra*, the actor-vists presented a dialogue between a character called William Shakespeare and Brad Patterson, supposed 'Arts Engagement Liaison at BP'. In the guise of Brad Patterson, the RSC were able to make the link between the theatre company, the oil industry and the targeted audience very clear:

> Giving is central to the fabric of BP's brand identity and so we are thrilled to sponsor 5 pounds tickets for 16–25-year-olds here at the Oil Shak—oops, I mean the Royal Shakespeare Company. Are there any 16–25-year-olds in the audience? Heyyyy!! [dabs] BP are committed to bringing the best of the UK's culture to a younger audience. We hope this can be the beginning of a beautiful friendship between BP and the next generation of artists, dreamers, and any budding oil men—or women!—in tonight. I like to think that if Shakespeare were alive today he'd be an oil man: he certainly knows his way around a geopolitical conflict![18]

That *Antony and Cleopatra* is partly set in Egypt is a salient fact and the RSC's website discusses the alliance between BP and the Sisi government.[19] However, by contrast with the intervention in 2012, which explicitly tied the RSC's performance to the Royal Shakespeare Company's season of plays, this later piece relies on the parodic portrayal of BP as one of the most significant long-term corporate investors in UK arts and culture.[20] Riffing on Macbeth's speech in *Macbeth* 5.5.18–22, the RSC's William Shakespeare makes the hard-hitting point that 'Your business plan scours and scalds our earth/ You emblacken our seas, unroot gentle trees,/ Untomorrow our days'.[21] Inasmuch as BP's Brad Patterson uses Shakespeare to try and influence the hearts and minds of the youthful theatregoers, the RSC makes their William Shakespeare disrupt the corporate pitch and leave the stage with the refrain that had been gaining momentum since 2012: 'Out damn logo!'[22]

BP's sponsorship of the arts (much like an oil slick itself) seeps out into other areas of the UK's cultural landscape. In September 2017, three actor-vists from the RSC showed how corporate finance is part of an interconnected network by invading the BP Lecture Theatre inside the British Museum (also sponsored by BP) just before Greg Doran, the then Royal Shakespeare Company's Artistic Director, was about to give a talk on the influence of the Roman poet Ovid on Shakespeare's plays and poems. As part of a campaign later described as 'Avid audience enjoys Ovid Protest', Performer 2 delivered 11 lines from Book 1 of Ovid's *Metamorphoses* describing earth's Iron Age degradation:

> And now the ground,
> which once—just like the sunlight and the air—
> had been a common good, one all could share,
> was marked and measured by the keen surveyor—
> he drew the long confines, the boundaries.
> Not only did men ask of earth its wealth,
> its harvest crops and foods that nourish us,
> they also delved into the bowels of earth:
> there they began to dig for what was hid
> deep underground beside the shades of Styx:
> the treasures that spur men to sacrilege.[23]

Performer 1 glossed the Ovidian quotation by bringing together the Royal Shakespeare Company's and the British

Museum's debt to BP. As in Stratford, the RSC ask the audience to look closely at BP's branding which surround them. 'Around the world', the RSC announced, 'BP stands for death and destruction—but here in the British Museum and in the [Royal Shakespeare Company] its logo is transformed into something friendly and benign, a partner for the arts'.[24] Working with the *Metamorphoses*' trope of transformation itself, recognised by ecocritical Shakespeareans as an extended example of 'biomorphic transmutation' between non-human and human modes of existence,[25] the RSC explained that

> BP is trying to purchase normality and acceptance, while going about its daily business of wrecking the very climate we all rely on for survival…don't be fooled by BP's attempts at shape-shifting—see it for what it really is, see its real impact on the world.[26]

Performer 3 ended this intervention with 9 lines from Book 8 of the *Metamorphoses*:

> Those roaring, rolling waters are quite used
> to bearing off stout trees and giant rocks.
> I've seen great stables standing on those shores
> and seen them carried off–livestock and all;
> against such surge, the bullocks' power failed,
> the helpless horses' speed did not avail.
> And when the snows along the mountain slopes
> have melted, this–my torrent–often swallows
> the bodies of young men in its wild whirlpools.[27]

In Ovid's poem, the river god Acheloüs delivers these lines as a warning to prevent Theseus from crossing the swollen river.

In the BP Lecture Theatre, Acheloüs' speech becomes part of a call to the audience to take action against climate catastrophe.

Between 2012 and 2017, the RSC based their actor-vism against the Royal Shakespeare Company's financial collaboration with BP in indoor site-specific adaptations of Shakespeare's works. Nine months later, and for their fiftieth performance, the RSC's increasingly ambitious campaign shifted from Shakespeare-focused interventions to one which took issue with the Royal Shakespeare Company's general enterprise by means of wide-ranging and immersive forms of creative protest. On 16 June 2018, the RSC returned to Stratford-Upon-Avon to 'pull off' their 'most ambitious intervention yet':

> Over 70 performers, ten performances, amazing special guests, a Shakespearean Insult Booth, a foyer invasion, a ceremonial performance about BP's abuses and guerrilla projections leave the RSC in no doubt about our feelings on their choice of oily sponsor.[28]

This five-hour Fossil Free Mischief Festival took place outside the Royal Shakespeare Theatre and mirrored the Royal Shakespeare Company's annual Mischief Festival at the Other Place, an event already primed for exploring twenty-first century issues of social justice, Shakespeare and education. The topic for 2018 addressed 'global questions of truth and freedom':[29] what better context for the RSC's ongoing push against BP's sponsorship of the arts? *BP or not BP?* recall that:

> [the] performance outside the Other Place aimed to highlight the hypocrisy of the RSC staging plays on these topics while partnering with BP. Under a giant BP logo

> with the words 'BP Also Sponsors', the performers held
> up phrases like 'Repressive Regimes', 'Violence and
> Corruption', and 'Colonialism'. They read out statements to
> show how BP is deeply complicit in the repression of protest
> around the world, through its partnerships with human-
> rights abusing governments—including Mexico and Turkey,
> the countries from which the Mischief Festival plays draw
> their stories. The performance included a call for the return
> of disappeared people in Mexico.[30]

Using BP's own marketing strategies and the co-option of art organisations against them, the RSC gained local and national publicity for their immediate project while simultaneously highlighting the global climate catastrophe. If BP use Shakespeare to make fossil fuel more palatable, as Evelyn O'Malley argues, *BP or not BP?*'s 'campaigning is made palatable through Shakespeare'.[31]

Instead of speaking from the stage, *BP or not BP?*'s Fossil Free Mischief Festival's players moved among their spectators and, as they returned to the Shakespearean text with which their protest began in 2012, literally brought them into actor-vism:

> Finally, Caliban from *The Tempest* summoned a dancing
> storm—with the help of the audience—to defeat BP.[32]

Here, Caliban's inclusive reclamation of *The Tempest* from Prospero's colonising practices makes an important statement about BP's culturally exploitative behaviours. In terms of the Royal Shakespeare Company's desire to link entertainment with fossil-fuel extraction, the RSC also showed that the theatre company didn't need the oil industry's money to fund the subsidised ticket scheme in the first place.

While the RSC's sustained, ultimately successful, portfolio of performances against the Royal Shakespeare Company's economic alliance with BP might be viewed as a niche dramatic genre confined to the realms of creative protest, it's worth placing their efforts alongside Theresa J. May's idea of 'ecodramaturgy', that is 'theater and performance making that puts ecological reciprocity and community at the center of its theatrical and thematic intent [...] carrying new frames for thinking about theater and new approaches and challenges to making theater'.[33] In their groundbreaking 2018 special issue of the journal *Shakespeare Bulletin* on 'Eco-Shakespeare and Performance' (and the source of my information about May's ecodramaturgy), Randall Martin and Evelyn O'Malley's editorial introduction discuss the RSC's Shakespeare in terms of the Royal Shakespeare Company's own attempts to produce environmentally aware theatre. Martin and O'Malley say:

> At the time of writing, the [Royal Shakespeare Company] has not dropped BP sponsorship, and it will be challenging to claim that any production is ecologically progressive while this remains the case—however much we might wish for the emancipatory potential of the aesthetic to challenge institutional policy from within a production.[34]

After seven years of focused activism using a range of dramaturgical techniques, we now know that the RSC made the interdependent relationship between the Royal Shakespeare Company and BP so visible that the Stratfordians had no other choice than to relinquish the fossil-fuel sponsorship. Far from being adjunct to mainstream theatre performance, the RSC deserves recognition as an ecodramatic theatre company in its own right. Beyond Stratford, the RSC's commitment to climate change, Shakespeare and performance as environmental activism cracks those troubling 'oil glasses' that render the fossil-fuel industries' activities almost imperceptible.

NOTES

1 Alejandra Borunda, 'We still don't know the full impacts of the BP oil spill, 10 years later', *National Geographic*, 20 April 2020. www.nationalgeographic.com/science/article/bp-oil-spill-still-dont-know-effects-decade-later para 3.
2 Cited in Brooks Hays, 'Deepwater Horizon spill has long-term effects on dolphins' immune systems', UPI, 18 February 2021. www.upi.com/Science_News/2021/02/18/Deepwater-Horizon-spill-has-long-term-effects-on-dolphins-immune-systems/6661613654563/ paras 1–2.
3 Heidi C.M. Scott, *Fuel: An EcoCritical History* (London: Bloomsbury, 2019), p. 178.
4 'About BP or not BP?', *BP or not BP?*, 2023. https://bp-or-notbp.org/about/ para 1. I'm extremely grateful to Jess Worth co-founder of BP or not BP? for taking the time to speak with me about the Reclaim Shakespeare Company's work.
5 For further information, see Stephen H. Grant, 'No Standard Oil Company? No Shakespeare Collection!' *Collation: Research and Exploration at the Folger*, 10 December 2019. www.folger.edu/blogs/collation/no-standard-oil/.
6 For further information about the RSC's initial aims and objectives, see 'Behind the curtains of the Reclaim Shakespeare Company', *BP or not BP?*, 28 August 2012. www.youtube.com/watch?v=QDzAW1C6DIg.
7 Matthew Taylor, 'Royal Shakespeare Company to end BP sponsorship deal', *The Guardian*, 2 October 2019. www.theguardian.com/stage/2019/oct/02/royal-shakespeare-company-to-end-bp-sponsorship-deal para 15.
8 'Protesters take to the stage at RSC over BP sponsorship', *BP or not BP?* 23 April 2012. https://bp-or-not-bp.org/2012/04/23/protesters-take-to-the-stage-at-rsc-over-bp-sponsorship/ para 6. See also the *Guerrilla Shakespeare Project* http://guerrillashakespeare.org/.
9 A copy of the script can be found on the RSC's website: https://bp-or-not-bp.org/action-is-eloquence/.

10 I am indebted to Lowell Duckert's use of this term. Lowell Duckert, *For All Waters: Finding Ourselves in Early Modern Wetscapes* (London: University of Minneapolis Press, 2017).

11 'The BP brand', BP. www.bp.com/en/global/corporate/who-we-are/our-brands/the-bp-brand.html.

12 'Helios at 20', BP. www.bp.com/en/global/corporate/news-and-insights/reimagining-energy/helios-at-20.html.

13 Clive Couldwell, 'BP at London 2012: Fuelling the future', *AV Magazine*, 14 March 2013. www.avinteractive.com/features/case-studies/bp-london-2012-14-03-2013/.

14 Susan Bennett, 'Sponsoring Shakespeare', in *Shakespeare's Cultural Capital: His Economic Impact from the Sixteenth to the Twenty-First Century*, edited by Dominic Shellard and Siobhan Keenan (London: Palgrave Macmillan, 2016), pp. 163–80, p. 163.

15 Clare Brennan, '*Twelfth Night: The Tempest* Review', *The Guardian*, 29 April 2012. www.theguardian.com/culture/2012/apr/29/twelfth-night-tempest-rsc-review para 1.

16 Caroline Spurgeon, *Shakespeare's Imagery and What It Tells Us* (Cambridge: Cambridge University Press, 1935), p. 24.

17 Steve Mentz, 'Toward a blue cultural studies: the sea, maritime culture, and early modern English literature', *Literature Compass*, 6/5 (2009): 998–1013, pp. 1001–2.

18 'Shakespeare *himself* just invaded the RSC stage in outrage at BP sponsorship', BP or not BP?, 27 April 2017. https://bp-or-not-bp.org/2017/04/27/shakespeare-himself-just-invaded-the-rsc-stage-in-outrage-at-bp-sponsorship/ para 15.

19 'BP, Egypt and sunken cities', *BP or not BP?* 16 May 2016. https://bp-or-not-bp.org/2016/05/16/bp-egypt-and-sunken-cities/.

20 'BP and leading UK cultural institutions extend partnerships for a further five years', BP, 28 July 2016. www.bp.com/en_gb/united-kingdom/home/news/press-releases/bp-and-leading-uk-cultural-institutions-extend-partnerships-for-a-further-five-years.html para 1.

21 'Shakespeare *himself* just invaded the RSC stage in outrage at BP sponsorship', *BP or not BP?*.

22 'Shakespeare *himself* just invaded the RSC stage in outrage at BP sponsorship', *BP or not BP?*.

23 'Avid audience enjoys Ovid protest', *BP or not BP?*, 29 September 2017. https://bp-or-not-bp.org/2017/09/29/avid-audience-enjoys-ovid-protest/.
24 'Avid audience enjoys Ovid protest', *BP or not BP?*
25 Randall Martin, *Shakespeare and Ecology* (Oxford: Oxford University Press, 2015), p. 29.
26 'Avid audience enjoys Ovid protest', *BP or not BP?*
27 'Avid audience enjoys Ovid protest', *BP or not BP?*
28 'Performances and Films', *BP or not BP?*, 16 June 2018. https://bp-or-not-bp.org/performances-and-films/#:~:text=Over%2070%20performers%2C%20ten%20performances,Out%2C%20damned%20logo! para 29.
29 'Mischief Festival Spring 2018', *RSC: Royal Shakespeare Company*. www.rsc.org.uk/mischief-festivals-past/mischief-festival-spring-2018.
30 'Fossil Free Mischief Festival comes to RSC's doorstep', *BP or not BP*, 17 June 2018. https://bp-or-not-bp.org/2018/06/17/fossil-free-mischief-festival-comes-to-rscs-doorstep/ para 8.
31 Evelyn O'Malley, *Weathering Shakespeare: Audiences and Open-Air Performance* (London: Bloomsbury, 2020), p. 189.
32 'Fossil Free Mischief Festival comes to RSC's doorstep', *BP or not BP*, 17 June 2018. https://bp-or-not-bp.org/2018/06/17/fossil-free-mischief-festival-comes-to-rscs-doorstep/ para 6.
33 Cited in Randall Martin and Evelyn O'Malley, 'Eco-Shakespeare in Performance: Introduction', *Shakespeare Bulletin*, 36.3 (2018): 377–90, p. 382.
34 Martin and O'Malley, 'Eco-Shakespeare in Performance: Introduction', p. 386.

Smoke

London's burning

Two

When I purchased a copy of The Clash's eponymous album on 8 April 1977 (its day of release), I had no idea that side one's closing anthem 'London's Burning'—a two-minute ten second testament to urban ennui[1]—had a loose connection to Shakespeare's *The Taming of the Shrew*. It's pretty clear from the track's alarum 'London's burning! London's burning!' that Joe Strummer and Mick Jones's lyrics echo the well-known English nursery rhyme of the same name:

> London's burning, London's burning
> Fetch the engine, fetch the engine
> Fire, fire! Fire, fire!
> Pour on water, pour on water.[2]

The association with Shakespeare's Elizabethan play takes a bit more teasing out. 'Still popular in schools today', Amelie Roper explains how the nursery rhyme

> is often sung in a round, with each singer starting after the previous one has sung one line of text. The words are often considered to be about the Great Fire of London. However, the earliest known notated version actually dates from 1580 and bears the words 'Scotland it burneth'.[3]

DOI: 10.4324/9780429326752-3

Next, Roper comments that the late-sixteenth-century 'song is alluded to in Shakespeare's *The Taming of the Shrew* Act 4 Scene 1 when Grumio asks Curtis to prepare a warm fire for guests':

Curtis: Who calls so coldly?
Grumio: A piece of ice. If thou doubt it, thou may'st slide from my shoulder to my heel, with no greater a run but my head and my neck. A fire, good Curtis.
Curtis: Is my master and his wife coming, Grumio?
Grumio: O ay, Curtis, ay; and therefore 'fire, fire; cast no water'. [4.1.15–16][4]

By comparison with The Clash's interest in the inflammatory cultural context of England's capital city in the late twentieth century, the premodern songs about London and Scotland offer literal scenes of conflagration which comment on their respective historical moments' social and, as I'll suggest in the following discussion, environmental politics. This section thus begins with an extended discussion of *The Taming of the Shrew*'s fire-filled dialogue before turning to fuel and its unburnt particles, smoke.

Before I continue, I want to make one thing very clear. I am not conflating the physical injury and ideological cruelty foreground in *The Taming of the Shrew*'s dramatization of courtship and marriage with other modes of harm. This play is all about domestic violence: Shakespeare's comedy asks critically aware audiences to interrogate the narrative.[5] With my book's general topic in view, I'm proposing that *The Taming of the Shrew*—a text so thoroughly absorbed in humankind's abusive treatment of its own species—provides ingress into thinking about Elizabethan England's contribution to current environmental concerns and attitudes to non-human interactions.

As part of the play's overarching plot concerned with the dampening of women's independence, the apparently slight moment of scene-setting at the beginning of Act 4 deftly develops Petruccio's telling metaphor in 2.1. Described in 'the persons of the play' as 'A gentleman of Verona', Petruccio arrives 'to wive it wealthily in Padua' (1.2.72) and ends up at the moneyed Baptista Minola's household to make a case for marrying Minola's eldest daughter:

> ... I tell you, father,
> I am as peremptory as she proud-minded,
> And where two raging fires meet together
> They do consume the thing that feeds their fury.
> Though little fire grows great with little wind,
> Yet extreme gusts will blow out fire and all.
> So I to her, and so she yields to me,
> For I am rough, and woo not like a babe. (2.1.128–35)

In this address to Katherine's father, Petruccio's phrase 'I am as peremptory as she proud-minded' makes his objective very plain: he will broker no debate with Baptista's daughter who is said to be 'intolerable curst, / And shrewd and froward so beyond measure' (1.2.85–6). Though Petruccio casts his rough wooing of Katherine in terms of fire-fighting tactics, his figurative language also engages with premodern humoralism's materiality.

One of the four elemental threads connected to humoral theories of the body, fiery concepts are commonly linked to sex. In the passion-propelled poem *Venus and Adonis*, for example, a verse recounting the goddess of love's predation of a mortal youth, Venus is described as being 'red and hot as coals of glowing fire' (35). A few hundred lines later, Venus uses the same figure to try and persuade Adonis to respond to her ardent pursuit. 'Affection', Venus proposes, 'is a coal that must be cooled, / Else, suffered, it will set the heart on fire' (387–8). *Venus and Adonis* has so much in common with *The Taming of the Shrew*'s disturbing outlook on the connections of power and sex that it could almost be subtitled 'the taming of the youth'. However, unlike the play's open-ended resolution, the poem shifts to the protracted death of the goddess' object of lust.

While indebted to similar tropes of passion, *The Taming of the Shrew* foregrounds humoral fire's production of yellow bile and its allied choleric temperament. In many ways, Shakespeare's Katherine Minola is a fine example of European culture's view of so-called choleric women:

An excess of the hot, dry emotion of choler, or yellow bile, produced an angry disposition. Choler is valuable in great warriors but in the domestic world of romantic comedy, anger—especially the anger of women—represents a social problem for Shakespeare's age, which calls for strong therapeutic intervention.

How to manage female anger is the central question of *The Taming of the Shrew*. Both protagonists, Kate and Petruchio, are identified as choleric by nature, and his behavior in the play is widely seen as eccentric and disruptive. Yet, it becomes Petruchio's job as husband to tame his shrewish wife and make her 'a Kate conformable as other household Kates'. [2.1.269–70][6]

With these infernal aspects of the play in mind, Grumio's humoral-infused dialogue with Curtis as they warm the house for their returning master and his new bride in 4.1—Grumio even describes himself as 'a little pot and soon hot' (4.1.5–6) in other words a small, angry person[7]—magnifies fire's positive attributes for physical well-being.

The Taming of the Shrew's Induction sets up its 'mock-elite smellscape' by 'burning sweet wood' (0.1.45).[8] It's important for the main plot that Katherine arrives at her new husband's dwelling in need of warmth, food and emotional sustenance and that Petruccio's home is equipped to cater for her needs. In most productions of the play, having clearly been subjected to an arduous 50-mile trip from Padua to Verona, Katherine takes the stage in a bedraggled form. From here on, Act 4 focuses on Petruccio's plan to 'kill [his] wife with kindness,/ And thus…curb her mad and headstrong humour' (4.1.188–9), a 'proper, fitting, appropriate'[9] brand of 'kindness', which involves starvation, sleep deprivation and a premodern form of gaslighting designed to challenge Katherine's perception of the world. Initially at odds with her new domestic environs, by the end of the play, Katherine's fiery reputation appears subdued.

The Taming of the Shrew's fifth and final act takes place at a banquet held at Lucentio's Paduan house. This occasion is designed to 'close [their] stomachs up/ After [their] great good cheer' and provide a venue for 'chat' (5.2.9–11). In keeping with comedy's generic criteria, Act 5 brings the plot's three heteronormative couples (Lucentio and Bianca; Hortensio and Widow; Petruccio and Katherine) together for a glimpse of their future married lives. Lucentio's promised conversation soon turns to the married couples' relationships. Widow's misunderstanding of Petruccio's views on the power dynamics in her own marriage swiftly turns to Widow's scrutiny of Petruccio's union with 'a shrew' (5.2.29). Katherine's subsequent retort to Widow ultimately becomes a spectator sport for their husbands and the rest of the male banqueters. Petruccio's and Hortensio's initial cheers of 'To her, Kate!' and 'To her, widow!' (5.2.34–5) incites bawdy banter. Led by Bianca, the present household's new mistress, the three women 'Exit' (s.d. 5.2.49) and leave their partners and the other male guests to place bets on the most obedient wife in a competition devised by Petruccio:

> Let's each one send unto his wife,
> And he whose wife is most obedient
> To come at first when he doth send for her
> Shall win the wager which we will propose. (5.2.67–70)

Only Katherine returns. In so doing, she secures her husband's gamble of a hundred crowns and is then sent to fetch the other wives who 'sit conferring by the parlour fire' (5.2.106). As with the start of Act 4, this invocation of domestic fire is significant. Bianca's and Widow's proximity to fire—that humoral tie to choleric temper—is linked to *The Taming of the Shrew*'s extended portrayal of Petruccio and Katherine's relationship which began with incendiary signs.

With everyone reassembled, Petruccio continues to demonstrate the success of his 'peremptory' approach to courtship by telling Katherine 'that cap of yours becomes you not. / Off with that bauble, throw it underfoot' (5.2.125–6). To both Widow's and Bianca's dismay, Katherine follows her husband's instruction. But that's not the end of Petruccio's performance of authority. He bids Katherine to 'tell these headstrong women/ What duty they do owe their lords and husbands' (5.2.134–5). In one of the most difficult Shakespearean episodes for twenty-first century feminist criticism,[10] Katherine delivers 44 lines to Widow and Bianca, which explains that an angry woman

> is like a fountain troubled,
> Muddy, ill-seeming, thick, bereft of beauty,
> And while it is so, none so dry or thirsty
> Will deign to sip or touch one drop of it.
> Thy husband is thy lord, thy life, thy keeper,
> Thy head, thy sovereign, one that cares for thee,
> And for thy maintenance commits his body
> To painful labour both by sea and land,
> To watch the night in storms, the day in cold,
> Whilst thou liest warm at home, secure and safe,
> And craves no other tribute at thy hands
> But love, fair looks, and true obedience,
> Too little payment for so great a debt. (5.2.146–58)

Like the cap she has just removed, Katherine's shifting images of 'painful labour' and domestic sanctuary recall her own journey into conjugal submission which gained traction in the preceding act. The speech ends with Katherine's advice to her fellow wives to 'place [their] hands below [their] husband's foot' (5.2.181) as she is now prepared to do. This emblem of women's subjection is followed by her husband's invitation 'Come on, and kiss me, Kate' (5.2.184). Once they've embraced, Petruccio instructs 'Come, Kate, we'll to bed' and the couple leave the stage (5.2.188). The play's final couplet, delivered by the remaining husbands, highlights their colleague's success in vanquishing 'a curst shrew' (5.2.192).

There's so much to question at the end of *The Taming of the Shrew* that it's hard to know where to begin. To take up and develop this section's interest in fire, fuel and smoke, I want to consider Katherine's point about women residing warm and safe at home under their husband's protection. If Act 4 shows the audience just how 'politicly' Petruccio's 'reign' (4.1.168) takes shape in his own home, then Act 5 provides a brief glimpse of Lucentio's and Bianca's marital life inside their dwelling. Ultimately, both houses envelop toxic scenes of domesticity that cannot be ignored. While Katherine's last speech asks women to appreciate their homes' shelter, the play provides insight into some of the ways patriarchal culture promotes her closing perspective. It turns out that Act 4's and Act 5's respective representations of fire's ability to keep women 'warm at home' can also cause great harm.

By comparison with the comedic onstage action, which brings the chimney into view as one of Falstaff's potential hiding places in *The Merry Wives of Windsor* 4.2.43, *The Taming of the Shrew*'s use of Elizabethan domestic architecture is embedded rather differently. In a play so concerned with social hierarchy and sexual politics, *The Taming of the Shrew*'s onstage fire in Petruccio's house and the offstage fire in the parlour of Lucentio's dwelling uphold these specific plot-driven interests. To intensify Katherine's social relocation from daughter to wife, film and theatre productions often depict Petruccio's household as an overwhelmingly masculine-coded environment. In Shakespeare's Globe's 2012 production, for example, Act 4 opens with a quintet of Petruccio's male servants singing the lascivious shanty 'The Cuckoo's Nest' surrounded by stag-horn themed furnishing (1:30:34–1:34:06).[11] Grumio's called-for heat is represented by a candle, a stage prop indicative of the open fires which commonly provided heat outward from a premodern building's centre up until the mid-sixteenth century.[12] The sort of raging domestic fire featured in Franco Zeffirelli's 1967 screen adaption, which allows Elizabeth Taylor's thoroughly unkempt Katherine to try and warm herself upon entering her new home, would clearly be a tricky scene to portray onstage (1:13:41–1:21:00). In any case, Zeffirelli's film cuts Grumio's and Curtis's key speeches at the start of 4.1 and the cinematic fire stands in for much of the excised text's exposition. In the Shakespeare's Globe's recording of their 2012 production, you can see traces of the candle's heat and smoke (1:34:42–1:42:05), fiery trails almost impossible to detect in live performance. By contrast with Act 4's emphasis on fire in general, the fact that we're told that Widow and Bianca convene around Lucentio's parlour fire suggests a more up-to-date technology involving chimneys and fireplaces.

Since G.W. Hoskins drew attention to the early modern English enterprise of the 'Great Rebuilding' in the 1950s, 'a revolution in the housing of a considerable part of the population' which took place between the mid-sixteenth and mid-seventeenth centuries,[13] twenty-first century studies have scrutinised numerous household differences afforded by these developments for the 'middling sort', notably 'the domestic arrangements of those below the ranks of the established gentry'.[14] With a narrative focused on the activities of middling households, Shakespeare's *The Taming of the Shrew* took shape during this period of 'Great Rebuilding' and what looks like mundane dramaturgy at first glance reaches out to contemporaneous material conditions of social and environmental remodelling. In these respects, performances of the play can emphasise how *The Taming of the Shrew*'s invocations of figurative and literal fire engage with the play's politics of control. Though diminutive, candles are potentially volatile flames harnessed in humankind's hands, a reminder provided by the Shakespeare's Globe's 2012 production when one of Petruccio's employees snuffs out the lit candles placed on the table around which the newly wed couple sit down to supper between 4.1.122 and 4.1.158 (1:41:43, 1:42:05). No doubt this action is for health and safety reasons rather than dramatic effect. But whatever the motive, there's only so much that humans can do to safely control fire's vital energy.

It's well-known that the original Globe Theatre was destroyed in 1613 by an errant spark from a small cannon fired during a performance of *All Is True* (*Henry VIII*) to herald 'a noble troop of strangers' (1.4.50sd-54).[15] Sir Henry Wotton's eye-witness account provides an animated description of how the theatre's destruction ran 'like a train' from a whiff of 'smoak':

> Now, to let matters of State sleep, I will entertain you at the present with what hath happened this week at the banks side. The Kings Players had a new Play, called *All is true*, representing some principall pieces of the raign of *Henry* 8, which was set forth with many extraordinary circumstances of Pomp and Majesty, even to the matting of the stage; the Knights of the Order, with their Georges and Garter, the Guards with their embroidered Coats, and the like: sufficient in truth within a while to make greatness very familiar, if not ridiculous. Now, King Henry making a Masque at the Cardinal, *Wolsey's* house, and certain Chambers being shot off at his entry, some of the paper, or other stuff wherewith one of them was stopped, did light on the thatch, where being thought at first but an idle smoak, and their eyes more

attentive to the show, it kindled inwardly, and ran round like a train, consuming within less then an hour the whole house to the very ground.

This was the fatal period of that virtuous fabrique, wherein yet nothing did perish, but wood and straw, and a few forsaken cloaks; only one man had his breeches set on fire, that would perhaps have broyled him, if he had not by the benefit of a provident wit put it out with bottle Ale. The rest when we meet.[16]

From a twentieth-first century perspective, Juan Carlos Skewes and Debbie Guerra explain how 'Fire is an event with a surface' and 'smoke, as a result of combustion, fills the [air] and new evanescent substances as fleeting surfaces are visible in its midst'.[17] Wotton's seventeenth-century account moves through a range of surfaces and 'stuff'—'the banks side', stage matting, 'embroidered Coats', paper, thatch, wood, straw, breeches—thus capturing the speed with which combustible materials interact with and transform each other into smaller 'fleeting surfaces'.

No human fatalities were recorded during this Jacobean theatrical disaster. However, devastating outbreaks of fire were a significant problem for Elizabethans. Suzannah Lipscomb's 2015 television documentary *Hidden Killers of the Tudor Home* features the everyday dangers caused by the design of domestic chimneys, specifically the problems of trying to achieve an effective 'draw' (the chimney's suction), understanding smoke's combustibility and the need to sweep chimneys regularly.[18] In Shakespeare's birthplace of Stratford-upon-Avon, for instance, during his lifetime 'there were three great fires in…1594, 1595, and 1614'.[19] While Katherine's final speech in *The Taming of the Shrew* makes an obvious point about some of the dangers faced by husbands' overseas and inland journeying, her comment about wives' domestic safety as they lie 'warm at home, secure and safe' underplays the household situation. Moreover, when viewed from the twenty-first-century English perspective, Katherine's concluding point about 'payment' and 'debt' takes on a much wider remit about the politics of privilege and what 'home', security and safety might mean on an environmental level.

I mentioned in this book's Introduction that *As You Like It*'s Forest of Arden is 'a crucial character in its own right'. For many ecocritics, that play's forest provides an important site for examining the rise of deforestation in premodern England alongside more positive developments 'such as the bio-dynamic farming and green rotations of corn and rye referenced in...1.1, 5.3'.[20] Yet Randall Martin's discussion of Mistress Quickly's reference to sea-coal in both *Henry IV Part Two* and *The Merry Wives of Windsor* (2.1.81, 1.4.8) and Mistress's Page's 'more expensive fuel preference at the conclusion' of the latter play's nuptial celebrations 'around a wood-burning "country-fire"' (5.5.219) shows how Shakespeare's Elizabethan texts are inscribed by England's growing and increasingly problematic fuel economy:

> Ancient English woodland and forests had been shrinking throughout the middle ages. By Henry VIII's time the pace began to accelerate. Worried about timber supplies for shipbuilding, the government took the first steps—largely ineffective—to manage depletions. Climactic and demographic pressures aggravated overexploitation, and by the 1590s caused a fuel crisis in south-east England and the country's first major environmental controversy. Similar to the threat of global warming temperatures today,

> the stresses on southern England woodland—at that time the country's most essential but finite natural resource—reached an ecological turning point...Wood shortages and soaring prices led domestic consumers to switch to coal—specifically, sea coal, which was mined near the surface in coastal areas of Newcastle and Scotland, and brought down England's eastern coast by barge.[21]

Martin's analysis of Mistress Quickly's and Page's respective nods to different kinds of domestic fuel flags up social difference. At the same time, the quotation above shows how a hierarchical geographical divide is implicated in premodern England's sea-coal supply chain: the north of England and Scotland serve the south.

Fossil fuel's extraction in late-sixteenth-century England took place away from the nation's main centres of coal consumption. And London consumed more coal than any other part of England. William M. Cavert states that 'The growing city of London, where so many people came during the long reign of Queen Elizabeth, soon used more of this coal than anywhere else'.[22] In the same decade as *The Taming of the Shrew*'s inception, William Cecil remarked how 'London and all other towns near the sea…are mostly driven to burn coal…for most of the woods are consumed'.[23] The affects were pungent, visible and enduring. Just over a hundred years later, John Evelyn's *Fumifugium.; Inconveniencie of the aer and smoak of London dissipated* described in some detail how smoke marked the city and its surfaces:

> [London] is here Eccplised with such a Cloud of Sulphure, as the Sun itself, which gives day to all the World besides, is hardly able to penetrate and impart it here; and the weary

> Traveller, at many Miles distance, sooner smells, then
> sees the City to which he repairs. This is that pernicious
> Smoake which sullyes all her Glory, superinducing a sooty
> Crust or furr upon all that it lights, spoyling the moveables,
> tarnishing the Plate, Gildings and Furniture, and corroding
> the very Iron-bars and hardest stones with those piercing
> and acrimonious Spirits which accompany its Sulphure; and
> executing more in one year, then expos'd to the pure Aer
> of the Country it could effect in some hundreds.... It is this
> horrid Smoake which obscures our Churches, and makes
> our Palaces look old, which fouls our Clothes, and corrupts
> the Waters, so as the very Rain, and refreshing Dews which
> fall in the several Seasons, precipitate this impure vapour,
> which, with its black and tenacious quality, spots and
> contaminates whatsoever is expos'd to it.... it is this which
> diffuses and spreads a Yellownesse upon our choycest
> Pictures and Hangings:[24]

'Pernicious' and 'horrid', Evelyn shows how the sulphurous condition of London's smoke taints everything in its wake, from the air to the water, indeed 'whatsoever is expos'd to it'.

Like the Elizabethan plays that Martin explores, *The Taming of the Shrew*'s specific focus on wealthy men and women's domestic life provides an opportunity for thinking about the implications of using wood and fossil fuels in the late sixteenth century and beyond. In *The Taming of the Shrew*, it's not too difficult to see that the further up the Elizabethan social hierarchy you go the more distant your relationship with fuel becomes. The comic business surrounding Grumio's and Curtis' preparation of the fire for their master and his new wife shows Petruccio's labouring community at work, from making the domestic fire to bemoaning the lack of the smokey by-product of 'link' (that is a torch) 'to colour Peter's hat' (4.1.114).²⁵ In many respects, the play's plot pretty much rests on Petruccio's servants' production of domestic fire—how else can the audience gauge the kind of household Katherine enters and watch how her husband's regime of 'taming' via starvation continues without the promise/denial of cooked meat?—and yet a key aspect of early modern fire is invisible. Of course, no one wants to see a stage full of smoke if it doesn't add anything to the performance. My interest here is to think about *The Taming of the Shrew* in its context in 1590 and what aspects of that context have been overlooked thus far. While Wotton's account of the Globe's 1613 ruin above and the early modern proverb 'no smoke without some fire'²⁶ connect cause and effect, *The Taming of the Shrew* features fireside gatherings without smoke, a rare occurrence in Elizabethan London.

Initially, London's smoke pollution wasn't caused by domestic fires. It turns out that the life-saving ale featured in Wotton's retelling of the Globe's fire has a link to London's contaminated air. In 1579 Elizabeth I took measures against the city's brewers (and 'brewing probably did consume close to half of London's coal') which mainly aimed to regulate the production of smoke in her immediate vicinity, especially 'when she "took her noble pleasure upon the water"'.²⁷ Seven years later, the city brewers were in trouble again as the Queen left the city 'because she was "greatly grieved and annoyed with the taste and smoke" of sea coal'.²⁸ As Cavert puts it,

> In these two episodes...Elizabeth's government established a principle that would remain at the heart of attempts to

restrict smoky industry during the seventeenth century: that coal smoke was a problem insofar as it affected the monarch's home and those spaces that were deemed essential to monarchical display.[29]

By the late 1590s, the distinction between industrial and domestic smoke production was less pronounced. Given the prominence of the pollution problem in London, it's not surprising some of Shakespeare's contemporary playwrights used smoke as a complex signifier of social difference:

In George Chapman, Ben Jonson, and John Marston's *Eastward Ho* (1605) coal smoke is invoked to define [a] social boundary by a character notable primarily for her desire to transcend it. Gertrude, the daughter of a rich goldsmith, wants to marry Sir Petronel Flash, whom she believes to be a gentleman, and thereby to become a lady and her parents' social superior. Escaping her family, their middling status, and her native environment combine in her plea to her betrothed: 'Sweet knight, as soon as ever we are married, take me to thy mercy out of this miserable City! Presently carry me out of the scent of Newcastle coal, and the hearing of Bow-bell' [...]. The smell of coal smoke, for her, defines London as effectively as the famous bells of St Mary-le-Bow. To leave the city's atmosphere is also to trade the mercantile world for the chivalrous values which she dreams will be upheld by her 'sweet knight'.[30]

The acrostic Argument prefacing Jonson's later single-authored city comedy *The Alchemist* (1612) foregrounds the flammable energy behind smoke and the enduring quality of these new 'fleeting surfaces':

> T he sickness hot, a master quit, for fear,
> H is house in town: and left one servant there.
> E ase him corrupted, and gave means to know
> A cheater, and his punk; who, now brought low,
> L eaving their narrow practice, were become
> C ozeners at large: and, only wanting some
> H ouse to set up, with him they here contract,
> E ach for a share, and all begin to act.
> M uch company they draw, and much abuse,
> I n casting figures, telling fortunes, news,
> S elling of flies, flat bawdry, with the stone:
> T ill it, and they, and all in fume [smoke] are gone.[31]

Here, Jonson's tropes of transmutation and transience alongside the author's self-conscious performance of textual agility function as plot summary and a subtle metatheatrical comment on that plot. In other words, while *The Alchemist*'s offstage explosion in 4.5.57–9 signals the destruction of the 'cozeners' enterprise, the ambivalent use of the pronouns in the Argument's final line suggests that the performance ('it') and the actors ('they') also disappear like 'fume'.

Shakespearean invocations of smoke are different than Jonson's. We already know from Wotton's account that *All Is True* used cannon fire to underpin the play's dramatization of pomp and ceremony. In a play likely to have been *The Taming of the Shrew*'s contemporary,[32] *Henry VI Part One* includes the stage direction 'Here they shoot [off chambers within]' (1.6.47 sd). In a wide-ranging discussion of the relationship between key theatrical performances of military topics and miasmic pollution in plague-ridden Elizabethan England, Chloe Preedy argues that 'Shakespeare interrogate[s] how gun-powder-powered warfare's perceptible impact on aerial environment might complement martial ambitions of conquest'.[33] The opening scene of *Titus Andronicus* links ceremony, battle and smoke:

> See, lord and father, how we have performed
> Our Roman rites, Alarbus' limbs are lopped
> And entrails feed the sacrificing fire,
> Whose smoke, like incense, doth perfume the sky.
> (1.1.142–45)

Lucius uses the smoke of 'the sacrificing fire' to help conjure a scene of post-war victory rather than visceral execution of their Goth prisoner of war. It is left to Aaron the Moor to expose Roman smoke as a sign of violence rather than veneration when he vows to protect his child 'Or some…will smoke for it in Rome' (4.2.110). Beyond ritual and war, Shakespearean smoke has a limited range of reference. Lucius' rhetorical stance in 1.1 chimes with his Uncle Marcus' unseemly poetics when he encounters Lavinia's terrible mutilation in 4.1. For different reasons, both Andronici men might be charged with using obfuscating verbal techniques such as the one mentioned in *King John* when that play's titular character accuses his opponents of deceitful speech as they 'shoot but calm words folded up in smoke' (2.1.229). Commenting on witty rather than political wordplay, *Love's Labour's Lost*'s Don Armado exclaims that Moth uses a 'Sweet smoke of rhetoric' (3.1.54) while the tragic Lucrece bespeaks of her own 'helpless smoke of words' (*The Rape of Lucrece* 1027). It's far more common for newly shed Shakespearean blood to smoke rather than steam (a surface I'll consider in the next section): the Sergeant in *Macbeth* speaks of 'brave Macbeth.../ Disdaining fortune, with his brandish'd steel steel /Which smoked with bloody execution' (1.2.16–18); the Second Gentleman 'with a bloody knife' in *The Historie of King Lear* announces Goneril's demise with the lines 'It's hot, it smokes/ It came even from the heart of ...Your lady, sir, your lady' (24.218sd-220); Antony in *Julius Caesar* refers to the 'purple hands' of Caesar's assassins 'which do reek and smoke' (3.1.159). When Tarquin makes his predatory nighttime moves towards Lucrece, 'the wind wars with his torch to make him stay, / And blows the smoke of it into his face' (*The Rape of Lucrece* 311–12). But *The Rape of Lucrece*'s literal alliance of fire and smoke is not a Shakespearean commonplace. And it's certainly not apparent in *The Taming of the Shrew*. Preedy suggests that Shakespeare's *Henry VI Part One*, *Henry IV Part One*, *Henry IV Part Two* and *Henry V* are 'sensitive to the contagious potential of an open-air theatre whose sulphuric fumes always threatened to drift out through the playhouse's unroofed centre and into the London atmosphere'.[34] Is it possible that Shakespearean texts as a whole are too aware of London's aerial pollution and Queen Elizabeth's displeasure to openly dwell on the topic?

While *The Taming of the Shrew* draws upon the labour and architecture of urban fuel consumption, its by-product smoke is left well alone. Nevertheless, the dramatization of domestic heat's importance in Acts 4 and 5 helps to raise questions about the politics of carbon that speak to its own time and our own. And if, as thinkers such as Simon Lewis and Mark Maslin suggest, the Anthropocene Epoch started in 1610,[35] then maybe *The Taming of the Shrew* turns our attention closer to England's influence on climate and the environment at the turn of the sixteenth century.

NOTES

1. The Clash, 'London's Burning'. https://youtu.be/weGEv_BhjNw.
2. 'London's Burning', *Timeless Children's Songs: For a New Generation of Music Lovers*. www.youtube.com/watch?v=5Atpbo3wOts.
3. Amelie Roper, 'London's Burning', *The British Library Music Blog*, 6 September 2016. https://blogs.bl.uk/music/2016/09/londons-burning.html. See also *The Norton Shakespeare*, p.196fn3.
4. Roper, 'London's Burning'.
5. For a detailed discussion of the play's troubling topic please read Emily Detmer, 'Civilizing subordination: Domestic violence and *The Taming of the Shrew*', *Shakespeare Quarterly*, 48.3 (1997): 273–94.
6. 'The unruly woman: The case of Katherine Minola', *National Library of Medicine, National Institutes of Health*. www.nlm.nih.gov/exhibition/shakespeare-and-the-four-humors/index.html#section3 paras 1–2.
7. *The Norton Shakespeare*, p. 204n4.1.
8. Chloe Preedy, *Aerial Environments on the Early Modern Stage: Theatres of the Air, 1576–1609* (Oxford: Oxford University Press, 2022), p. 220.
9. OED 1b.
10. In a review of the Royal Shakespeare Company's 2012 production Maddy Costa asks if *The Taming of the Shrew* is 'An exercise in misogyny— or a love story about a man liberating a woman?' Maddy Costa, 'The Taming of the Shrew: 'This is not a woman being crushed': *The Guardian*, 17 January 2012. www.theguardian.com/stage/2012/jan/17/taming-of-the-shrew-rsc.
11. Toby Frow and Ross MacGibbon (dirs), *The Taming of the Shrew* (Globe on Screen, 2012), *Drama Online* recording. www-dramaonlinelibrary-com.ezproxy.lancs.ac.uk/video?docid=do-9781350997530&tocid=do-9781350997530_4598380313001.
12. See further Tara Hamling and Catherine Richardson, *A Day at Home in Early Modern England: Material Culture and Domestic Life, 1500–1700* (London: Yale University Press, 2017), p. 71ff.
13. W.G. Hoskins, 'The Rebuilding of Rural England', 1570–1640', *Past and Present*, 4.1 (1953): 44–59, p. 44.
14. Hamling and Richardson, *A Day at Home in Early Modern England*, p. 5.
15. *The Norton Shakespeare*, p. 3119.
16. 'Sir Henry Wotton on *All Is True (Henry VII)* and the Burning of the Globe (1613)', *The Norton Shakespeare*, pp. 3310–11.

17 Juan Carlos Skewes and Debbie Guerra, 'Air, Smoke and Fumes in Aymara and Mapuche Rituals', in *Surfaces: Transformations of Body, Materials and Earth*, edited by Mike Anusas, and Cristián Simonetti (London: Routledge, 2021), pp. 29–45, p. 36, p. 31.

18 Suzannah Lipscomb, *Hidden Killers of the Tudor Home* (London: Modern Television, 20 January 2015). www.youtube.com/watch?v=5zSyjyLA WWM 13.24–24.00.

19 Stanley Wells, 'Fires in Stratford-upon-Avon', *Oxford Reference* (Oxford: Oxford University Press, 2015) www-oxfordreference-com.ezproxy.lancs.ac.uk/view/10.1093/acref/9780198708735.001.0001/acref-9780198708735-e-1034?rskey=25pwtl&result=1.

20 Randall Martin, 'Shakespeare, ecology, and environmental concerns', *Shakespeare and Beyond: The Folger Shakespeare Library*, 18 April 2017 https://shakespeareandbeyond.folger.edu/2017/04/18/shakespeare-ecology-environmental-earth-day/ para 6.

21 Randall Martin, *Shakespeare's Ecology* (Oxford: Oxford University Press, 2015), pp. 2–3.

22 William. M. Cavert, *The Smoke of London: Energy and Environment in the Early Modern City* (Cambridge: Cambridge University Press, 2016), p. 17.

23 Cited in Ken Hiltner, *What Else Is Pastoral: Renaissance Literature and the Environment* (London: Cornell University Press, 2011), p. 98.

24 John Evelyn, *Fumifugium.; Inconveniencie of the aer and smoak of London dissipated* (London, 1661), p. 6. See further Emily Cockayne, *Hubbub: Filth, Noise and Stench in England 1660–1770*, new edition (London: Yale University Press, 2021), p. 152.

25 *The Norton Shakespeare*, p. 206n7.

26 Morris Palmer Tilley, *A Dictionary of the Proverbs in England in the Sixteenth and Seventeenth Centuries* (Ann Arbor: University of Michigan Press, 1950), S569.

27 Cavert, *The Smoke of London*, p. 25.

28 Cavert, *The Smoke of London*, p. 47.

29 Cavert, *The Smoke of London*, p. 48.

30 Cavert, *The Smoke of London*, pp. 197–8.

31 Ben Jonson, *The Alchemist*, edited by Elizabeth Cook, second edition (London: A & C Black, 1991), p. 27. All quotations are from this edition.

32 See the plays' respective essays and textual notes by Jean. E. Howard in *The Norton Shakespeare*, p. 159, p. 166, p. 465, p. 473.
33 Chloe Kathleen Preedy, 'The Smoke of War: From *Tamberlaine* to *Henry V*', *Shakespeare*, 15.2 (2019): 152–75, p. 169.
34 Preedy, 'The Smoke of War', p. 171.
35 Philip John Usher, *Exterranean: Extraction in the Humanist Anthropocene* (New York: Fordham University Press, 2019), p. 12.

Sky

Unfirming the firmament

Three

5 May 2021. Nothing could prepare me for the joyful feeling of walking the mile or so from my flat to my dentist's surgery on such a glorious day. I'd received my first Covid-19 vaccination a fortnight earlier and England was twelve days away from Step 3 of the government's 4-Step roadmap for lifting the lockdown. (The fact that I had a dental appointment in the first place was very welcome news by comparison with the dearth of other medical treatments currently available.) I stopped on Skerton Bridge crossing Lancaster's River Lune and took a photo to try and capture an almost-forgotten feeling of being outdoors and in sync with the sky above. The global pandemic's interruption of air-travel and its allied contrails helped fashion the clichéd—but nonetheless stirring—clear Wedgewood-like effects above me as the fair-weather cumulus clouds skimmed the skyline comprised of Lancastrian building and trees.

Twentieth- and twenty-first century thinkers such as James J. Gibson and Tim Ingold have theorised the kinds of sensations I experienced on my May-day walk. A few decades after the publication of Gibson's *The Ecological Approach to Visual Perception* (1979), which posited that 'the surface is where most of the action is',[1] Ingold reviews his predecessor's work on humankind's complex relationship

DOI: 10.4324/9780429326752-4

with the sky. In the wake of Gibson's work, Ingold compares both the physical world and the atmosphere as material conditions which:

> can exist only in relation to the forms of life that inhabit it. It is a world that exists not in and of itself, but as the ambience of its inhabitants. Though no less real than the physical world, the environment is not a reality *of* objects or bodies in space but reality *for* the beings that make a living there. Thus conceived, the environment—Gibson argues—'is better described in terms of a *medium* [air, water], *substances* [rock, gravel, sand, soil, mud, wood, concrete] and the *surfaces* that separate them' (1979: 16).[2]

In this generally structuralist view, the division between 'medium' and 'substance'—Gibson's surface—shares the following basic 'properties': 'a relatively persistent layout, a degree of resistance to deformation and disintegration, a distinctive shape, and a characteristically non-homogeneous texture'.[3] Ingold then explains how Gibson's surfaces are 'recognised' via a 'characteristic' interplay of texture and light. No visible texture equals no visible surface. No visible surface equals an empty void. 'The perception of the sky', argues Ingold, 'offers a case in point'.[4] Rather than leaving the sky as a surface-less expanse and a largely redundant aspect of Gibson's understanding of being-in-the-world, Ingold turns to environmental components (air, water, fire, earth) to argue that 'it is in the medium—and not on the surface, as Gibson thought—that 'most of the action is'.[5] While both writers value humankind's sensorial engagements with their surroundings, Gibson's predominant interest in sight gives way to Ingold's approach, which is largely informed by phenomenological interactions, the body's relationship with 'the incessant movements of wind and weather'.⁶ For Ingold, touch matters. In some ways, his idea of 'weather-world',⁷ a phrase designed to describe the atmosphere's all-embracing impact on human selfhoods, has more in common with the entangled sensibilities of pre-Cartesian Eurocentric thought. So how might Shakespeare's skies and clouds help us to think about weather-world on a sunny day in May 2021 and vice-versa?

Four hundred years before Gibson's and Ingold's respective theories on the agency of surfaces and media, Shakespeare's writings economically encapsulate existing ideas on some of the affinities between human selfhoods and a sky so obviously free from aircraft and artificial light pollution that's almost impossible to imagine. In premodern England, as Sophie Chiari tells us, the sky's significance was such that it

> remained ominous and threatening while astrological attempts to read one's destiny in the stars (well before William Herschel's observations) or in the clouds (when Luke Howard's classification was not even conceivable) long continued to flourish. At any rate, in the Geneva Bible, Shakespeare's contemporaries could find plenty of stories telling of the erratic weather caused by God's wrath and were used to reading the changes in the English firmament in the light of the Scriptures. Most of them still thought that, because of men's ingrained tendency to sin, the rain poured down from the heavens, threatening to engulf the peasants' yearly harvests, if not humanity at large, and flashes of lightning cracked in angry skies, punishing human hubris.⁸

Chiari shows how the meaning of the Christian premodern sky is dependent on a set of religious and cultural codes that still have a bearing on my own secular sense of meteorological activity. When I recall the gentle flow of Lancaster's River Lune on 5 May 2021, it's hard to remember the local havoc caused six years earlier by Storm Desmond (such a homely name for a weather event that has 'the potential to cause disruption or damage which could result in an amber or red warning')[9] and the destruction triggered by the same river breaching its defences. On 5 December 2015, a height of 8.17 metres was recorded (the maximum normal height is 3.60 metres)[10] and with a speed of around 1,750 m^3 per second, the River Lune had 'the highest flow ever recorded on an English river'.[11]

Primarily, my memory of domestic incidents like this one triggers just how much climate privilege the UK currently has in relation to so many parts of our world. As I redraft this section, for example, on 30 August 2022, the BBC reported that 'One-third of Pakistan has been completely submerged by historic flooding'. In the words of Pakistan's climate minister Sherry Rehman: 'It's all one big ocean, there's no dry land to pump the water out… [this is a] crisis of unimaginable proportions'.[12] By 9 September 2022, Fatima Bhutto's powerful opinion piece for the *Guardian* announced 'Today, Pakistan, the world's fifth-most-populous country, is fighting for its survival. Those who don't die from the floods risk death by starvation—yet you've probably heard little about the devastation'.[13] In the UK, 11 days of national mourning following the death of nation's monarch on 9 September 2022 placed restrictions on most news channels about anything other than the state's memorialisation of the erstwhile Queen. Specific details about Pakistan's heating up by 'the dreaded 2.2F', 'erratic monsoon rains', glaciers 'melting at a rate never seen before', how 'Sindh, the southernmost province, received 464% more rain over the last few weeks than the 30-year average for the period' and that 'Ninety per cent of crops in "that area" have been damaged' support Bhutto's point that 'This is a tragedy of nightmarish proportions'.[14] 'Forget solidarity', Bhutto warns, 'the global south will not survive this century without climate justice.'[15] The very least I can do is to keep in mind that the small part of north-west England I inhabit is ultimately connected to South Asia and the rest of the world not just ecologically but also intellectually. The fact that I ever think I'm not part of a global economy is an effect helped along by the kind of news coverage I'm exposed to and, of course, the kind of wider society I'm part of. Reading and thinking about Shakespeare's plays and poems transhistorically *and* glocally might help to think more fully about living in weather-worlds.

Premodernity's desire for biblical guidance alongside astrological forecasts and almanacs is completely explicable. Before national tracking systems like the UK Government's Environmental Agency and regular broadcasts of weather forecasts, it's easy to see how England's inhabitants gained reassurance and guidance from augurs, portents and scriptures when dealing with devasting weather conditions. But looking back at earlier ideas of the sky and its components also shows how this often-localised aspect of earthly dwelling at large is culturally and politically controlled.

Part of a European humoral economy enmeshing human-kind and the elements I discuss throughout this book, the premodern sky was at once legible and incomprehensible. While various kinds of cosmological and religious sign systems were developed, as with any kind of non-verbal and verbal languages, their meanings were often contested. And we need to remember that the English vernacular itself was in a state of flux following the social, political and theological upheavals of the Henrician Reformation in the 1530s. To begin with, and as Chiari's quotation above suggests, there are some differences between the predominantly secular sense of 'sky', 'The region of the atmosphere and outer space seen from the earth in which the sun, moon, stars and clouds appear', and the broadly sacred idea of 'firmament', 'The arch or vault of heaven overhead',[16] defined in the well-known opening lines of Genesis 1. On the second day of creation, as the post-Reformation 1560 Geneva Bible puts it:

> Again God said, Let there be a firmament in the midst of the waters, and let it separate the waters from the waters. Then God made the firmament, and separated the waters, which were under the firmament, from the waters which were above the firmament; and it was so. And God called the firmament Heaven. So the Evening and the morning were the second day. God said again, Let the waters under the heaven be gathered into one place, and let the dry land appear; and it was so. And God called the dry land, Earth, and he called the gathering together of the waters, Seas; and God saw that it was good.[17]

Sixteenth-century England thus envisions a terrestrial globe under a canopy which divides the earth's waters from the heaven's. Church fathers and classical philosophers argued about its precise nature, but in any case this interface called firmament 'for scriptural reasons...had to be simultaneously in contact with both the waters above and those below it'.[18] Though arborescent concepts such as the Great Chain of Being (E.M.W. Tillyard's phrase for the vertical order of God's creations, from lowly stones to the prime mover himself) prevails, the curved Christian covering now called firmament works in tandem with secular geocentric and heliocentric nests of crystalline spheres with either the earth or—more controversially—the sun respectively at its core.

A great deal has been written about the seismic cultural shifts eventually engendered by the 1543 publication of Nicolaus Copernicus' heliocentric work *On the Revolutions of the Celestial Spheres* in combination with the empirical evidence enabled by the invention of Galileo Galilei's astronomical telescope in 1609–12. It wasn't until February–March 1616 that the Catholic Church censured Copernicanism.[19] Protestant England, by comparison, responded with far less caution. Though 'Luther too despised Copernicus and all that he represented so the reception was no better in the continental Protestant world…England was really the only safe place to discuss these new ideas'.[20] Even so, late-sixteenth-century books such as Thomas Blundeville's *Exercises, Containing Sixe Treatises* (1594) (on arithmetic, cosmography, terrestrial and celestial globes, Petrus Plancius' world map, John Blagrave's astrolabe and the principles of navigation), which, the title tells us, 'are verie necessarie to be read and learned of all yoong Gentlemen that have not bene exercised in such disciplines', remained underpinned by biblical celestial architecture. Blundeville's sky, for instance, is divided into three hierarchical spherical regions. In the lowest section nearest to humankind, clouds, rains and dews are formed. Frosts, snow, ice and hail are made in the middle section while lightning, fire drakes and blazing stars occur in the upper band. Divided from these lower three spaces by either ether or a fifth element called quintessence (the celestial substance latent in all terrestrial things), God's heavenly realm looms above all.[21] Indebted at the outset to Claudius Clavius' work on Theodosius' *Spherics* and a translation of Albrecht Durer's *Geometrie* (1535) by a friend whose name Blundeville 'conceale[s] at his owne earnest intreatie', Blundeville's work is primarily a domestication of ideas designed to 'profite [his] countrey'.[22] Bound up with this Christian-influenced Anglicised view of the sky is an all-too-apparent objective to serve Elizabethan colonial expansionism.

Other forms of sky-related knowledge show off their political stance in different ways. Informed by Claudius Ptolemy's highly influential *Almagest*, the Greek-Alexandrian catalogue of 48 constellations, European celestial globes, some highly ornate and complexly mechanical, chart the positions of the stars with as much critical and creative interest as the maps focused on lands and seas. Ten years after Gerard Mercator constructed his table globe in 1541, for instance, he made a celestial globe which Gerhard Emmoser adapted in 1579 to fashion a remarkable clockwork celestial globe for the Holy Roman Emperor Rudolf II. With winged Pegasus transporting the spherical astronomical chart,[23] Emmoser's exquisite allegorical etchings on gilded silver forge typical humanist links between classical mythology and mid-sixteenth-century state-of-the-art technology.

Celestial globes like these connect human imaginations with the social worlds those imaginations inhabit. And while famous globes like Mercator's or Emmoser's bear the name of individual European men, their dependence on others' knowledge and expertise is largely obscured. Contingent upon a global matrix of intellectual expertise and constructed out of material hewn from the earth (silver, brass, steel, wood), these celestial globes should be testaments to the collaborative yet largely anonymous labour, which make these wonderful objects depicting wondrous perspectives manifest. I'll have more to say about materiality per se in section Ten's discussion of 'Stage: Disposable Globes'. What interests me for now is the celestial's globe mirror-image view of the sky that shows the planets and stars *above* the earth's surface while simultaneously placing human perspective *outside* that view. From their clearly earthbound position, the viewer is both within and without the celestial globe, a bifurcated stance which captures enduring tensions related to human subjugation and authority.

As I see it, the question of humankind's place in and power over premodern and modern ecologies is a puzzle that Shakespeare's works confront. My aim in this part of the book is to focus on the ideas of sky and firmament in general circulation in the late sixteenth and early seventeenth centuries; to consider broad understandings of 'This most excellent canopy the air' (*Hamlet* 2.2.290–91) and to think about key onstage dramatisations of those understandings.

If you type 'firmament' into the search function of *Open-Source Shakespeare*, this word meaning 'arch or vault of heaven, in which the clouds or stars appear'[24] occurs once in six individual plays. With its etymological roots in the Latin term *firmus* (firm), four tragedies—*Titus Andronicus* (5.3.17), *Richard II* (2.4.19–20), *Julius Caesar* (3.1.62) and *King Lear* (1.2. 121)—show male figures (Saturninus, the Earl of Salisbury, Caesar and Edmund, respectively) referring to either the sun or stars in the firmament as a quick comment on their own or someone else's stable/unstable character. In these contexts, the nature of the firmament itself, this solid overhead vault, is not in question. The treatment of the word 'firmament' in two plays in this sextet stand out from the rest: another tragedy *Hamlet* and the late-play *The Winter's Tale*.

Anne-Valérie Dulac points out that '*Hamlet*'s skyscape is essentially overcast, and the many thick clouds hanging over Shakespeare's Denmark are in no small part responsible for the tragedy's gloomy atmosphere'.[25] It's not altogether surprising then that the philosophical and ever-garrulous prince turns his attention aloft to ponder his precarious emotional and political circumstances once he's learned from his father's ghost that Claudius, his uncle/step-father/ King, is responsible for King Hamlet's murder. In conversation with his childhood friends, Rosencrantz and Guildenstern, who have been asked by King Claudius to find out what is wrong with his nephew/step-son/subject, the second Quarto's Hamlet tells them:

> I have of late—but wherefore I know not—lost all my mirth, forgone all custom of exercise; and indeed, it goes so heavily with my disposition that this goodly frame, the earth, seems to me a sterile promontory. This most excellent canopy the air, look you, this brave o'erhanging [firmament],[26] this majestical roof fretted with golden fire—why, it appears no other thing to me than a foul and pestilent congregation of vapours. (2.2.287–93)

Unlike the later Folio edition, which leaves a vague reference to skyward matter, the Second Quarto changes the noun 'o'erhanging' into an adjective and inserts firmament as the noun. The textual shift encourages a spiritual rather than secular interpretation, a Quarto-quibble, which is useful for thinking about Shakespeare's skies and surfaces in two ways.

Firstly, and in the wake of humorally—and religiously—inflected texts such as the fulsomely titled William Fulke's *A goodly gallery with a most pleasant prospect, into the garden of naturall contemplation, to behold the naturall causes of all kynde of Meteors, as wel fyery and ayery, as watry and earthly, of whiche sort be blasing sterres, shooting starres, flames in the ayre etc. tho[n]der, lightning, earthquakes,* etc. rayne dewe, snowe, cloudes, springes etc. stones, metalles, earthes etc. to the glory of God, and the profit of his creaturs (1563), a book forging all kinds of connections including links between the human brain and the sky,[27] Hamlet makes comparisons between his melancholy demeanour and the 'o'erhanging firmament', which seems to him like 'a foul and pestilent congregation of vapours'. Secondly, Hamlet summarises premodern Europe's elemental views of earth and its 'canopy the air' (in effect the two kinds of terrestrial and celestial globes discussed above) to which he adds 'a man' as a 'quintessence of dust' (2.2.294–8), in other words a wrecked example of the celestial stuff (that so-called fifth element) contained in everything. In a rhetorical mode so typical of this university-educated Prince, Hamlet at once shows: his character's erudition and precarity as King Hamlet's usurped heir-apparent; late-sixteenth-century cosmological debates; the interconnected nature of humankind with its surroundings.

Even though their dramaturgical and social status isn't that of Hamlet's, another son in *The Winter's Tale*'s gets straight to the matter of surfaces in a way that Denmark's Prince doesn't. In Act 3's extraordinary scene of infant conveyance, coastal storm and a stage direction ('Exit, pursued by a bear' 3.3.57sd) that's as well-known as a Hamletian soliloquy, the Old Shepherd's son (usually known as Clown) enters *The Winter's Tale* for the first time with a description of the sky's/firmament's slippery surficial and semantic boundaries and the limits of human of perception itself:

> I have seen two such sights, by sea and by land! But I am not to say it is a sea, for it is now the sky. Betwixt the firmament and it you cannot thrust a bodkin's point. (3.3.79–81)

For some, these lines might be evidence of the character's uninformed social status. In performance, for example, the Old Shepherd's offspring might demonstrate bewilderment, a sentiment connecting this figure with *The Winter's Tale* similarly bewildering plot. Just after these lines, Clown tells his father how he has seen a spectacular storm at sea, 'how it chafes, how it rages, how it takes up the shore' (3.3.83–4). But the 'point' of Clown's wonder is the ship caught up in the terrifying weather, 'O, the most piteous cry of the poor souls! Sometimes to see 'em, and not to see 'em' (3.3.84–6). The Clown swiftly turns from the seascape to the sea's side and the 'bear' who 'tore out [a man's] shoulder bone, how [the man] cried to me for help, and said his name was Antigonus, a nobleman!' (3.3.89–90). In this way, the audience learns of the ursine chase's grisly outcome—and that Clown didn't try and help. The scene ends with Clown saying he'll go and see if the bear has finished eating 'the gentleman' and 'If there be any left of him' (3.3.116–19). Act 3 then ends. Act 4 takes up the plot some 16 years later.

Sharing vocabulary with Chiari's explanation above, the juxtaposition of 'sky' and 'firmament' in Clown's speech succinctly engages with premodern England's understanding of humankind's integrated earthly and celestial environments. In so doing, Clown recalls Genesis' idea of creation, which rests upon the mindful division of waters under and above the firmament.

By contrast with Hamlet's richly textured prosody, in just three sentences the Old Shepherd's son captures the seeming fusion of maritime with aerial space *and* the cultural complexity of the extra-terrestrial canopy. Though clearly separated by class, as representations of a younger generation both Hamlet and the Old Shepherd's son dramatise how knowledge of the sky was variously questioned and understood.

Composed of a variety of verbal and non-verbal signs, a play is made up of a range of ideas dispersed through its own culture and those who watch and think about the play in future times. So far, I've drawn upon a few specific examples to give a sense of how premodern English artisans, authors and playwrights engage with sky/firmament. If premodern English culture's use of these nouns betray a slippery etymology, then its view of clouds is perhaps even more tricky. As Hamlet and Polonius' conversation about whether a 'yonder cloud' is 'in shape of a camel', a weasel or a whale (3.2.345–51) and the 2021 Australian news item with the headline 'Mum "creeped out" after spotting Shakespeare's face with goatee in cloud' suggest,[28] premodern and contemporary Anglo-Antipodean societies have an interest in seeing clouds as nouns and finding meaning in these elemental things.

By contrast with 'the fixèd stars of heaven' (*Richard II*, 2.4.9) set like jewels in the crystal spheres and arranged into constellations of recognisable forms, clouds present basic challenges. While the strenuous efforts of nineteenth-century Europeans to classify 'cloud physiognomy'[29] shows how difficult the task is, sixteenth-century writers like Fulke in the 'The fourth booke of watry impressions' from his *A Goodly Gallery* tries to categorise clouds in terms of: their height above the earth; whether they contain rain or if they're 'empty'; colours (redde fyre and yealowysh, 'whyte', 'black', 'blew' and 'grene').[30] Though it's an overtly Christian treatise (a quotation from Psalm 148 appears on the title page), Fulke's book makes some effort to provide empirical evidence and to bring a range of 'common opinion' and the more specialist views of 'geometrical demonstrations' together.[31]

But it's Leonardo da Vinci and his capacity for describing all kinds of visual forms who makes the most remarkable observations about moisture-laden shapes in his fifteenth-century notebooks. Da Vinci divides 'visible bodies' into 'two kinds: those 'without shape or any distinct or definite extremities' and those with a 'surface [that] defines and distinguishes the shape'.[32] While he doesn't mention clouds directly, they clearly fit into da Vinci's definition of bodies 'without surface':

> those bodies which are thin or rather liquid, and which readily melt into and mingle with other thin bodies, as mud with water, mist or smoke with air or the element of air with fire, and other similar things which are mingled with the bodies near to them, whence by this intermingling their boundaries become confused and imperceptible, which for reason they find themselves *without surface*, because they enter into each other's bodies, and consequently *such bodies are said to be without surface*. (my emphasis)[33]

Related ideas are evident in Shakespeare. A few years after Hamlet and Polonius' onstage dialogue, Shakespeare's cloud conversations continue in *Antony and Cleopatra* with Antony's views on shape-shifting identities:

> Sometimes we see a cloud that's dragonish,
> A vapour sometime like a bear or lion,
> A towered citadel, a pendent rock,
> A forkèd mountain, or blue promontory
> With trees upon't that nod unto the world
> And mock our eyes with air. Thou hast seen these signs;
> They are black vesper's pageants.
>
> ...
>
> That which is now a horse, even with a thought
> The rack [dislimns], and makes it indistinct,
> As water is in water. (4.15.2–11)[34]

Like *Hamlet*, this later tragedy engages with a cultural climate of cloud gazing and cultural consensus while linking Antony's meditation on cloudscapes to his own emotional state.³⁵ If Philo tells us at the start of the play that Antony's desire for Cleopatra 'O'erflows the measure. Those his goodly eyes, / That o'er the files and musters of the war/ Have glowed like plated Mars, now bend, now turn' (1.1.1–4), Antony's own view of his failing commitment to Rome now comes to the fore. He says to his attendant and fellow Roman Eros 'now thy captain is/ Even such a body. Here I am Antony, /Yet cannot hold this visible shape' (4.15.12–14). The vocabulary used to describe these beguiling overhead forms—vapour, air, rack—bespeak the liquid transience of meteorological matter which anticipates Prospero's proclamation in *The Tempest* that 'the great globe itself ... shall dissolve' and 'Leave not a rack behind' (4.1.153–6). Unlike the later play's apocalyptic image, however, Antony's thoughts are focused on a variety of clouds with the appearance of beasts, cityscapes and landscapes that 'nod unto the world' and the dissolution of corporeal bodies which 'dislimn' just like 'water... in water'. For Rhodri Lewis:

> The clinching term is 'dislimns'. There is an obvious pun here, but as Shakespeare knew well, and as is detailed in Nicholas Hilliard's *Arte of Limning*, to limn is to portray or to paint, usually in watercolour or some other kind of wash...In a brilliant inversion of the way in which the topos was commonly used, Shakespeare has Antony imaginatively transfigure the clouds not to create something new or to show off his inventive talent, but as an index of the way in which his formerly distinct self image has been *un*painted from within.³⁶

To extend Lewis' perceptive analysis a bit further, I'll add that Antony's cloud gazing shares common ground with da Vinci's painterly discussion of intermingled earthly substances and bodies without surfaces. But as with the insertion of the religiously inflected word 'firmament' into *Hamlet*'s Second Quarto, Antony's speech simultaneously ties these skyborne substances to the liturgical circumstances of 'black vespers'. To be sure, Antony's striking observation that the cloud's shape diffuses like 'water into water' is another Shakespearean echo of God's division of the waters of the earth from the waters of the heavens. This is a specific kind of chaos allied with the world's creation rather than its end and a reminder, perhaps, that in the beginning humankind did not (and does not?) take centre stage.

Given the Book of Genesis' patriarchal order of Adam and Eve's entry into the world, it's telling that—so far— I've only referenced fictional and historical premodern men commenting on humankind's relationships to skies, firmaments and clouds. Where are the women like Hypatia, the Alexandrian woman now known for her editing of Ptolemy's *Almagest*? We only have to look at the Shakespearean plays I've discussed to see that women's voices are largely absent from premodern discussions of skies. And it can't be a coincidence that Sonnet 14's adoption of astronomical knowledge is addressed to the young man. It's not that Shakespeare's women are wholly ignorant of celestial topics. Julia in *The Two Gentlemen of Verona* (2.7.74) and Helena in *All's Well That Ends Well* (2.5.71), for example, make integrated remarks about stars plotting destinies. Juliet asks that Romeo should become a constellation when she dies, that the night should 'Take him and cut him out in little stars' (3.2.22). Imogen compares herself to an astronomer who is 'learn'd indeed' if they 'knew the stars as' well as she knew her husband's handwriting (3.2.27–8). But where are the Shakespearean women who think skyward in as much detail as Hamlet, Antony or Clown? This kind of gender divide is an obvious indicator of European men's desire to quite literally master the heavens.

A closer look at Emmoser's fabulous globe alongside Mercator's is helpful here. Emmoser's globe's 'has extra constellations', Clare Vincent remarks, 'and they both turn out to be unclothed ladies, probably made specifically to please the emperor'.[37] Vincent doesn't say anything further about the globe's depictions of these additional constellations who take the form of naked women. Markedly, they turn out to be groups of stars named after named after Egyptian queens regnant from the same Ptolemeic Dynasty that includes Cleopatra, Shakespeare's sovereign figure. On the one hand, some might suggest that Emmoser's globe is deifying these powerful Alexandrian women by 'cut[ting them] out in little stars'. On the other hand, since the publication of Edward Said's groundbreaking thesis on *Orientalism* (1978), it's impossible to read these etchings of unclothed Egyptian queens—women rulers reduced to salacious snapshots for powerful men—as anything other than premodern Christian European examples of the West's exoticisation of the East. Emmoser's celestial globe is not concerned with understanding the sky: it's all about authorizing Rudolf II's command of it.

Like Emmoser's luxurious object, Shakespeare's works are part of a European period approaching Galilean invention and the destabilising effects of 'the ocular proof' (*Othello* 3.3.365). With the physical design of the firmament under telescopic scrutiny and its hierarchical view of earth and sky in question, Shakespeare's texts help to show how sacred and secular beliefs are managed and how some voices demand, take even, more attention than others. Hamlet and Antony use the sky to say something of their emotional experiences. They are staged examples of living in Ingold's idea of weather-world. Clown observes a devasting scene unfolding before him—and does nothing to respond to Antigonus' pleas for help and save any human life except his own. Rather than seeing Shakespeare's works as character-driven exemplars, the plays and the poems show how power works. Even if we inhabit different parts of the earth, we need to feel part of a global sky.

NOTES

1. James J. Gibson, *The Ecological Approach to Visual Perception: Classic Edition* (London: Psychology Press, 2015), p. 19.
2. Tim Ingold, 'Earth, sky, wind, and weather', *The Journal of the Royal Anthropological Institute*, 13 (2007): 19–38, p. 25.
3. Ingold, 'Earth, Sky, wind, and weather', p. 26.
4. Ingold, 'Earth, sky, wind, and weather', p. 26.
5. Ingold, 'Earth, Sky, wind, and weather', p. 28.
6. Ingold, 'Earth, Sky, wind, and weather', p. 34.
7. Ingold, 'Earth, Sky, wind, and weather', pp. 32–8. See also Tim Ingold, 'Weather-world', *The Life of Lines* (London: Routledge, 2015), pp. 69–72.
8. Sophie Chiari, *Shakespeare's Representation of Weather, Climate and Environment: The Early Modern Fated Sky* (Edinburgh: Edinburgh University Press, 2019), p. 1.
9. The UK Storm Centre. www.metoffice.gov.uk/weather/warnings-and-advice/uk-storm-centre/index.
10. 'The River Lune Level at Caton', *Gov UK*. https://check-for-flooding.service.gov.uk/station/5029.
11. 'Lancaster Flood Risk Management Scheme'. *The Flood Hub*. https://thefloodhub.co.uk/lancaster/.
12. Leo Sands, 'Pakistan floods: One third of country is under water—minister', BBC News. 30 August 2022. www.bbc.co.uk/news/world-europe-62712301, paras 1–3.

13 Fatima Bhutto, 'The west is ignoring Pakistan's super-floods. Heed this warning: tomorrow it will be you', *The Guardian*, 9 September 2022. www.theguardian.com/commentisfree/2022/sep/08/pakistan-flo ods-climate-crisis para 1.
14 Bhutto, 'The west is ignoring Pakistan's super-floods', paras 1–3.
15 Bhutto, 'The west is ignoring Pakistan's super-floods', para 15.
16 OED 'sky', n.1; 'firmament' n.1a.
17 1: 6–10 www.genevabible.org/geneva.html.
18 Edward Grant, *Planets, Stars and Orbs: The Medieval Cosmos, 1200 to 1687* (Cambridge: Cambridge University Press, 1996), p. 102.
19 Maurice A. Finocchiaro, '400 years ago the Catholic Church prohibited Copernicanism', *Origins: Current Events in Historical Perspective*, The Ohio State University, February 2016. https://origins.osu.edu/milestones/ february-2016-400-years-ago-catholic-church-prohibited-copernican ism?language_content_entity=en.
20 Michael Rowen-Robinson, 'Shakespeare's Astronomy', Gresham College Lecture, 30 November 2016. https://brewminate.com/shakespeares- astronomy/ paras 3–4.
21 My understanding of Blundeville's work is indebted to Chloe Preedy, *Aerial Environments on the Early Modern Stage: Theatres of the Air, 1576–1609* (Oxford: Oxford University Press, 2022), pp. 13–15.
22 Thomas Blundeville, *M. Blundevile his Exercises, Containing Sixe Treatises* (1594), A3r–A4r.
23 See *The Royal Museums Greenwich Collection*, www.rmg.co.uk/stor ies/topics/explore-stunning-16th-century-table-globe-3d and The Metropolitan Museum, www.metmuseum.org/art/collection/search/ 193606.
24 OED 1.A.
25 Anne-Valérie Dulac, 'The "Mass and Charge" of Clouds in *Hamlet* (4.4.47)', in *William Shakespeare: Hamlet, Prince of Denmark: Perspectives Critiques*, edited by Denis Lagae-Devoldère, Mickaël Popelard and Guillaume Winter (Paris: Ellipses, 2022), pp. 25–41, p. 25. I'd like to thank Anne-Valérie Dulac for sending me a copy of her essay.
26 See *The Norton Shakespeare*, p. 1725fn1.
27 Preedy, *Aerial Environments on the Early Modern Stage*, pp. 67–74. For a discussion of Fulke and *Hamlet*, see Carla Mazzio, 'The history of air: *Hamlet*

and the trouble with instruments', *South Central Review*, 26: 1–2 (2009): 153–96.

28 Emilia Shovelin, 'Mum creeped out after spotting Shakespeare's face with goatee in cloud', *The Sun*, 14 December 2021. www.thesun.co.uk/news/16862056/stunning-photo-shakespeare-in-clouds-thought-she-was-hallucinating/.

29 Lorraine Daston, 'Cloud Physiognomy, *Representations*, 135 (2016): 45–71.

30 Fulke, *A Goodly Gallery*, 46v–48v.

31 Fulke, *A Goodly Gallery*, 47r–47v.

32 Edward MacCurdy (ed), *The Notebooks of Leonardo da Vinci* (New York: George Braziller, 1955), pp. 986–7.

33 MacCurdy, *The Notebooks of Leonardo da Vinci*, p. 987.

34 *The Norton Shakespeare* gives 'distains' (without note) rather than 'dislimes' as found in the First Folio. For this reason, and following The Arden Shakespeare's *Antony and Cleopatra* edited by John Wilders (London: Bloomsbury, 1995), I have cited 'dislimns'.

35 Seee further Rhodri Lewis, 'Shakespeare's clouds and the image made by chance', *Essays in Criticism*, 62.1 (2012): 1–24.

36 Lewis, 'Shakespeare's clouds and the image made by chance', pp. 7–8.

37 Clare Vincent, 'The celestial globe that features mythology, science, and technology, *The Metropolitan Museum of Art*, 26 November 2013 www.metmuseum.org/perspectives/videos/2013/11/celestial-globe-art-explained 0:50-0:58.

Steam

Under pressure

Four

Before Covid-19 entered our lives, my daily encounters with steam—mainly in the kitchen or in the shower—generally passed me by. Now, steam is something I think about every time I board public transport or enter a crowd-filled space. My need to put on a face covering with prescription eyewear has meant trying all kinds of hints, tips and tricks to deal with my exhaled breath which ends up as a distorting film of condensation on my spectacles' lenses.[1] And when I become more anxious, the condensation becomes worse. Without a doubt, this mundane experience has increased my awareness of the obviously vital but usually invisible processes of respiration at individual and communal levels.

At the start of the global pandemic, media discussion about the need for adequate ventilation in work and leisure spaces was generally subordinate to the role surfaces played in the transmission of the disease.[2] Even after the World Health Organization (WHO) published online guidance on 10 January 2020, which 'recommend[ed] droplet and contact precautions when caring for patients, and airborne precautions for aerosol generating procedures conducted by health workers',[3] interest in the surface-level transmission of the corona virus remained.

Example 1. Just over a year after the WHO guidelines first appeared, and with a specific focus on retail environments,

DOI: 10.4324/9780429326752-5

Dyani Lewis' article for the British scientific journal *Nature* observed how 'it's easier to clean surfaces than improve ventilation—especially in the winter—and consumers have come to expect disinfection protocols'.[4] To put it another way, it's less effort—and cheaper—to show surfaces rather than air being cleaned. And vigorous cleaning also puts on a good show: this is 'hygiene theatre'.[5]

Example 2. Twenty-four months after the WHO's advice was published, debates about poor air quality in UK primary and secondary schools and the amount of government support needed to provide air filters to reduce the aerial transmission of Covid-19 among children continued.[6] Indeed, the nation's public health protocols remained unclear. In the words of Claire Horwell, a professor of geohealth at Durham University, UK:

> It is now accepted unequivocally that the virus responsible for Covid-19 is airborne, travelling metres on an infected person's breath within liquid microdroplets and aerosols. One of the most important measures to prevent transmission of an airborne respiratory virus is the use of masks. Any face covering is better than nothing, but in the UK the government hasn't informed people about the most effective protection, or ensured that they have access to it.
>
> Nearly two years into the pandemic, official guidance remains confusing. The government still discourages non-healthcare workers from obtaining PPE, including certified respiratory protection, stating that it is used in a limited number of industrial and healthcare settings.[7]

It seems that my spectacles' steamy surface is a sure sign of the need for improved protection. According to Horwell, 'as a general rule of thumb, if you feel air blowing into your eyes and your glasses fog rapidly…then contaminated air can also get in'.[8] In this specific twenty-first century miasmic moment, steam is a marker of humankind's precarity; the respiratory systems required for life can also lead to death.

Shakespeare's *Venus and Adonis* explores related concerns about air, breath and human vulnerability that connects premodern worries to those raised by the global pandemic of 2020. It's something of a commonplace to mention that this poem, Shakespeare's earliest publication, was produced while London's theatres were closed during an outbreak of the plague. Still, this late-sixteenth-century context is relevant for my discussion of the poet's sole use of the word 'steam' which occurs in the early part of this tragi-comic narrative (at line 63) about Venus' desire for Adonis, a mortal youth who would rather hunt than engage in a sexual encounter with the goddess of love. For anyone familiar with the poem's classical precedent (the story is most well-known from Ovid's *Metamorphoses*), the fact that Adonis dies in the pursuit of a boar rather than respond to Venus' amorous advances is no surprise: Adonis is always in danger. In what follows, I suggest that the poem's fleeting glance at steam's evanescent surface engages with the fear of unseen airborne threat in the sixteenth and twenty-first centuries.

As we saw at the end of the previous section 'Smoke', there are some figurative correspondences between Shakespeare's representations of blood's heat and traces of unburnt fuel. In *Julius Caesar* (3.1.159), *Macbeth* (1.2.18) and *The Tragedy of King Lear* (5.3.195sd-197), for instance, Shakespeare's depictions of human bloodshed are emphasised by fiery remnants not moisture-laden trails. Katherine Craik's understanding of 'the flexibility of the early modern humoral system' is helpful for summarising the relationship between blood and smoke in these three plays:

> Flashes of fire, or choleric outbursts, could... be either the reprehensible signs of a personality trait shared among the numberless masses, or might confirm that one's blood degree was of the highest order.⁹

Julius Caesar, *Macbeth* and *The Tragedy of King Lear* link blood and fire around aristocratic usurpation, murder and suicide by sword or knife. Shakespeare's smoking blood thus bespeaks human culpability and high social rank. Like these plays, Shakespeare's *Venus and Adonis* is concerned with the death of a human subject by a wound. However, the bloody breach of Adonis' body is caused by an animal's tusk in the depths of a forest, not blade-wielding humans enmeshed in statecraft. And in the context of the poem's interests in emotions, embodiment and transformation, steam—the stuff of heat and water—is an apt short-lived surface for the poem to invoke.

At *Venus and Adonis*' tragic conclusion, we encounter the youth's corpse from the goddess' perspective:

> ...from their dark beds once more leaps her eyes,
> And, being opened, threw unwilling light
> Upon the wide wound that the boar had trenched
> In his soft flank, whose wonted lily-white
> With purple tears that his wound wept was drenched.
>   No flower was nigh, no grass, herb, leaf, or weed,
>   But stole his blood and seemed with him to bleed.
>   (1050–6)

As Adonis' blood floods the surrounding vegetation, Venus' thoughts turn to her own attempts to ensnare the mortal. In so doing, she imagines that 'If' the boar 'did see his face, why then, I know/ He thought to kiss him, and hath killed him so' (1109–10). That's it, she says,

> thus was Adonis slain;
> He ran upon the boar with his sharp spear,
> Who did not whet his teeth at him again,
> But by a kiss thought to persuade him there,
>   And, nuzzling in his flank, the loving swine
>   Sheathed unaware the tusk in his soft groin.
> Had I been toothed like him, I must confess
> With kissing him I should have killed him first;
> But he is dead, and never did he bless
> My youth with his, the more am I accursed. (1111–20)

Simultaneously providing a motive for the boar's destruction of Adonis while aligning 'her own lust ... with that of the boar',[10] Venus considers how she and the animal pursued the youth. And there is more than grief-stricken musing in the goddess' comparison. Dympna Callaghan points out how 'Venus' identity is essentially non-human, both bestial *and* immortal'.[11] Goddess and boar, non-humans both, survive the poem. In *Venus and Adonis'* interspecies love triangle, only the human form disappears. Like the pant that defined him at the start of the poem, the youth's body 'melted like a vapour from her sight, / And in his blood that on the ground lay spilled, / A purple flower sprung up, chequered with white' (1166–68). While the flower's appearance initially reminds Venus of Adonis' white body speckled with blood, 'the new-sprung' (1171) flower's smell is suggestive of the youth's breath. *Venus and Adonis* is thus framed by respiratory images which are a feature throughout the poem.

By modern standards, and as the quotation from the poem above suggests, *Venus and Adonis* is a steamy story.[12] This kind of Elizabethan verse is sometimes called an epyllion, a short-form epic which commonly covers erotic material.[13] First and foremost, *Venus and Adonis* is a tale about sex and the strategies employed by the goddess of love to trap a beautiful young man who prefers chasing animals with his presumably male friends—though it's never really clear who or what Adonis wants. I'm inclined to agree with the point that '*Venus and Adonis* is a poem about one-sided desire and that not only does Adonis not desire Venus, he doesn't really desire anyone or anything'.[14] In the wake of such critical analyses, and by comparison with the hyper-sexualised goddess, Adonis' apparently asexual standpoint is foreground.[15] Drawing attention to the association between Venus and the poem's use of 'gustatory metaphors', Wayne A. Rebhorn shows how

> Adonis really is a morsel for Venus to devour; the metaphors of eating not only suggest lust but also imply the swallowing up of Adonis' self that giving in to Venus' love entails. Adonis does indeed face danger that Venus will 'draw his lips' rich treasure dry' (552), a phrase that hints at vampire-like horrors, at loss of breath and vital fluid, perhaps even loss of soul.[16]

Indeed, there's something of the vampiric femme-fatale about Venus' predation of Adonis from the start of the poem when 'she seizeth on his sweating palm' and 'calls it balm', which is 'Earth's sovereign salve to do a goddess good' (25–8). At the very least, there's an extractive quality about Venus' craving for Adonis' perspiration. But behind this post-gothic reading of Shakespeare's Venus lies a far more premodern attitude allied with classical ideas of pneuma:

> The meanings of pneuma include breeze and breath/respiration but stretch to spirit, inspiration, and eventually to ghost, as in what used to be called the Holy Ghost, but is now more often referred to as the Holy Spirit.[17]

While it's hard to see the Christian aspects of this highly material poem, *Venus and Adonis*' engagement with various concepts of vaporous exchange bring Shakespeare's poem into this pneumatic domain. Though overlooked by Rebhorn, Shakespeare's sole use of 'steam' in *Venus and Adonis* is also part of the consumptive dynamic between goddess and mortal. Forty lines or so after grabbing Adonis' sweaty hand, Venus' desire for the mortal youth is captured in the narrator's description of how he 'breatheth in her face' and 'She feedeth on the steam as on a prey' (62–3). However, as Adonis' breath is taken by the goddess at the beginning of the poem (as his blood will be purloined by the plants in which his dead body lies at the end), Shakespeare's adaptation of Ovid's myth goes beyond the wholly interpersonal dimensions of the myth.

By contrast with a play, a poem's narrator can address a reader directly and mindfully guide them into a specific textual arena. On the one hand, and as Coleridge realised, *Venus and Adonis* draws the reader into an intimate study 'of every outward look and act' and 'of the flux and reflux of the mind in all its subtlest thoughts and feelings'.[18] On the other hand, Shakespeare's non-dramatic verse steers its audience, and the Ovidian episode from the *Metamorphoses* Book 10 on which it's based, into a much wider cultural context. The poem's titular protagonists are placed in a verdant, sparsely populated micro-climate, a context which had some resonance in Elizabethan England. One year before Shakespeare's poem was first published in England, Abraham Fraunce's Ovidian-inspired text *The Countess of Pembroke's Ivychurch* (1592) provided a cosmological commentary on the myth:

> Now, for Venus her love to Adonis, and lamentation for his death: by Adonis, is meant the sunne, by Venus, the upper hemisphere of the earth [...] by the boare, winter: by the death of Adonis, the absence of the sunne for the sixe wintrie monethes; all which time, the earth lamenteth: Adonis is wounded in those parts, which are the instruments of propagation: for, in winter the son seemeth impotent, and the earth barren: neither that being able to get, nor this to beare either fruite or flowres.[19]

Though the allegory is by no means original to the author, the proximity of Fraunce's mythopoesis to Shakespeare's *Venus and Adonis* enables us to think about the 1593 poem in ecological terms. Instead of focusing on Venus' frustrated desire, Fraunce's interpretation unlocks the poem's wider setting. Shakespeare's *Venus and Adonis* draws on the Latin poem's 'biomorphic transmutations',[20] Randall Martin's cogent phrase used to define the *Metamorphoses*' non-human/human transformations. In both the classical and the vernacular verses, the tragic outcome is the same: killed by the wild boar, Adonis is transformed into a flower. But after reading Fraunce, we can see that one of the key differences between the classical source and Shakespeare's version is the English poem's extended interest in the weather.

Part of the *Metamorphoses*' seamless weave of over 220 myths, Ovid's version of Venus and Adonis (along with those of Ganymede, Hyacinthus, Cerastae and Propoetides, Pygmalion and Myrrha) is embedded in Orpheus' lengthy lament for the death of his beloved Eurydice: the tale has its own narrative tangent as the goddess recounts Atalanta and Hippomenes' mythic history to the youth. Orpheus begins his song on a sun-drenched, grassy plain but shade arrives, quite literally, as trees move in to cover the grieving poet: some trees even have their own stories to tell. In this way, Ovid's *Metamorphoses* integrates non-human and human agencies into the storytelling process itself. Like its classical predecessor, one of the English author's favourite sources, Shakespeare's *Venus and Adonis* is interested in fashioning a specific scenario. Modern printed editions of the poem are usually accompanied by Shakespeare's dedication to the nineteen-year-old 'Right Honorable Henry Wriothesley, Earl of Southampton, and Baron of Titchfield' which explains that 'Only, if your honour seem but pleased, I account myself highly praised, and vow to take advantage of all idle hours till I have honoured you with some graver labour'.[21] Using the commonplace premodern convention of feigned ignorance, the dedication sets up a professional yet intimate relationship between the newly published poet and the young courtier for this vernacular Ovidian verse focused on human-like but not exclusively human forms.

Shakespeare's *Venus and Adonis* takes place in an Anglicised climate shaped by diurnal patterning. The poem's plot follows the sun's daily cycle from morning (as we've seen) to its 'midday heat' (177) to sunset (529–34) to darkness (811–16) and the following morning's sunrise (853–8). The 'bonnet' with which Adonis 'hides his angry brow' (339) and Venus' poignant memory of how 'The wind would blow [the bonnet] off and, being gone, Play with his locks' (1087–90) places the youth in the same sartorial orbit as the poem's young dedicatee. But it's the poem's realisation of the landscape that emphasises its domesticated environment. For some, in its invocation of 'blue-veined violets' (125), a 'primrose bank'(151), 'brook' (162) and 'hedge' (1094), the location seems far more reminiscent of 'the Forest of Arden in Warwickshire [rather] than a grove in Ancient Greece'.[22] Beyond Venus, Adonis and the boar, the poem's supporting cast of horses (30ff), hares (674,679,697), foxes (675), sheep (685), hounds (686,692), deer (676, 689) along with a soundscape featuring a huntsman's 'halloo' (973) offers a predictable English hunting scene.

In a premodern 'geo-humoral' context, Mary Floyd Wilson's concept describing a 'regionally framed humoralism',[23] Adonis' behaviour might be partially explained by the day's various moisture levels. The poem begins just after dawn's dew point, a time of day when the air temperature is cool enough to become saturated with water vapour:

> Even as the sun with purple-coloured face
> Had ta'en his last leave of the weeping morn,
> Rose-cheeked Adonis hied him to the chase.
> Hunting he loved, but love he laughed to scorn.
>   Sick-thoughted Venus makes amain unto him,
>   And like a bold-faced suitor 'gins to woo him. (1–6)

Venus and Adonis' exploration of love in the form of a 'sick-thoughted' goddess pursuing a Venus-averse 'rose-cheek'd [mortal] youth' starts, then, in a damp, cool temperate place which immerses its titular pair in a recognisable domestic environment. In some ways, and as part of the poem's Englishness, the evocation of a moist comparatively cold morning—typical of an English spring or early summer day—is aligned with premodern Hippocratic views that 'the putative cold wetness of Englishmen is continuous with the climate of Northern Europe'.[24] In this case, English temperaments are defined as phlegmatic, a term standing in for a range of attributes including dullness, lethargy and a predilection for emotional inconstancy due to the phlegmatic's exceptionally porous body, making it vulnerable to changes in the weather.[25] Is Adonis a version of an English phlegmatic? He's fairly dull and lethargic where Venus is concerned (she calls him a 'lifeless picture' (211)), but I've always found it tricky to know if when 'offers he to give her what she did crave./ ...He winks, and turns his lips another way' (88–90) is some evidence of Adonis' capricious attitude. Theories play out differently in practice, and Shakespeare's textual microclimate is inscribed with a range of contradictory 'somatic ecological'[26] positions that, put very simply, are different than our own. Hippocrates isn't the only classical authority behind premodern geohumoralism. Aristotelian theories, for example, posit that a cold climate *generated* internal heat.[27] And in this case, Adonis' heat in combination with Northern Europe's chilly temperatures might account for the poem's vapoury images.

In the poem's timeframe, Adonis' temperature evidently responds to the shifts in the sun's journey east to west. When he cries out that 'The sun doth burn [his] face' (186), for example, Venus tries to make herself into an alternative micro-climate:

> I'll sigh celestial breath, whose gentle wind
> Shall cool the heat of this descending sun.

> I'll make a shadow for thee of my hairs;
> If they burn too, I'll quench them with my tears.
> (189–92)

It doesn't work. But Venus' emphasis here on her own 'celestial breath' takes us back to the opening of the poem when she snatches Adonis from his horse and wrestles the youth to the ground. Bearing in mind that the narrator tells us that Adonis is 'forced to content' (61), that is 'acquiesce',[28] it's not obvious to me if his 'panting' (62) is due to physical exertion or anxiety induced by abduction. While the narrator describes the youth's exhalation via the matter-of-fact term 'steam', like the later reference to her own breath in the quotation above, Venus 'calls' Adonis' highly visible respiration 'heavenly moisture, air of grace' (64). To say that she puts a positive spin on her violent behaviour is an understatement.

Venus's attraction for the living Adonis is clearly motivated by sexual desire. But if we now follow the goddess' interest in Adonis' breath a bit further into the narrative, we find another striking remark that widens the poem's respirational perspective. Believing Venus has dropped dead at his rebukes, Adonis tries to resuscitate her in what amounts to a parodic version of the familiar poetic device of the blazon: 'He wrings her nose, he strikes her on the cheeks, / He bends her fingers, holds her pulses hard;/ He chafes her lips' (475–7). In spite of trying 'a thousand ways' (477) to bring her round, Adonis finally resorts to kissing her until she wakes. An ecstatic Venus revives and addressing Adonis' lips she announces:

> Long may they kiss each other, for this cure!
> O, never let their crimson liveries wear,
> And as they last, their verdure still endure,

> To drive infection from the dangerous year,
> That the star-gazers, having writ on death,
> May say the plague is banished by thy breath! (505–10)

It's no surprise that Adonis' lips are red, but Venus' mention of their vegetal qualities is an unexpected addition. As well as underscoring Adonis' youth, the lips' greenish hue is suggestive of the plants used throughout premodern Europe to prevent infection.[29] We've already seen that Venus thinks of Adonis' sweat as goddess 'balm'. With this later point that his breath prevents the plague, Adonis is in danger of becoming a kind of first-aid kit.

Forty lines or so later, the narrator's view of Venus' vapours is quite different than Venus' opinion of Adonis'. After agreeing to kiss the goddess goodnight, the youth quickly finds himself assailed once more as the goddess seeks to 'draw his lips' rich treasure dry' (552). At this, the narrator describes how 'having felt the sweetness of the spoil, / With blindfold fury she begins to forage, / Her face doth reek and smoke, her blood doth boil' (553–5). If Adonis' (anxious?) breath is likened to steam at the start of the poem, it's important to note that Venus' 'face doth reek', an intransitive verb that can also mean 'To emit or give off vapour or steam, especially under the influence of heat'.[30] Around the time of *Venus and Adonis'* publication, 'reek' was also starting to indicate 'unwholesome odour or fume'.[31] Adonis' clearly wholesome breath alongside Venus's steaming body offers a noteworthy sexual politics of steam. But there is more to say about her association with reeking smoke. Venus' insistence on physical intimacy and Adonis' refutation of close bodily contact chimes with contemporaneous fears about infection. In a discussion of 'How new diseases were investigated in the 16th century', Vivian Nutton explains that:

> Britain lagged behind most of Europe. It had no health boards and almost no town physicians before 1600. It was not until 1518 that arrangements were made for doing more than cleaning the streets and occasionally segregating the sick. There were no plague hospitals, or lazaretti, such as were found in Florence or Venice, and administration was left to individual parishes. How effective the continental system was at combatting epidemic disease is debatable.[32]

With its plague-riven historical context and the poem's own interest in respiratory infection, Shakespeare's airborne goddess is as much of a miasmic threat as she is a sexual predator.

In our own precarious time, faced with the ongoing risks of life-threating viruses and the collapse of earth's climate, we can learn a few things from Shakespeare's unusual invocation of the word 'steam' in *Venus and Adonis*. To begin with the obvious. Like my own experiences with Covid-related condensation, steamy breath makes humankind's being in the world manifest. As Tim Ingold states:

> The process of respiration, by which air is taken in by organisms from the medium and in turn surrendered to it, is fundamental to all life…For it is in the nature of living beings themselves that, by way of their own processes of respiration, of breathing in and out, they bind the medium with substances in forging their own growth and movement through the world. And in this growth and movement they contribute to its ever-evolving weave.[33]

As I've discussed in this section, in part *Venus and Adonis* is organised around breath, from Adonis's exhalation in line 62 to Venus' inhalation of the flower's fragrance in line 1171. If we follow that shift from the youth's panting to the goddess' sniff, as in its Ovidian predecessor, we can see that Shakespeare's poem is interested in the kind of 'ever-evolving weave' Ingold describes. And yet, I must mark that while *Venus and Adonis'* human protagonist dies, other forms of life continue.

Example 3. 'After two winters of Covid anguish, one would be forgiven for viewing the shortening of days with a sense of trepidation. It would not be entirely misplaced'.[34] As of September 2022, the UK's ability to overcome Covid-19 looks unpromising. I want to believe we mortals have a future on this planet. However, our failure to successfully address global pandemics could mean that, like Adonis, humankind disappears from view.

NOTES

1 See further Claire Horwell, 'Ministers know which masks provide the best Covid protection—why not tell the UK public?', *The Guardian*, 30 December 2021. www.theguardian.com/commentisfree/2021/dec/30/masks-best-covid-protection-ffp2-ffp3:

> In England, it became law to wear a face covering on public transport on 15 June 2020 and a second law, covering public indoor areas, such as retail spaces and places of worship, came into effect five weeks later. Both laws were revoked on 18 July 2021, on the eve of "freedom day". As a response to the threat of Omicron, a new law enforcing face coverings on public transport and in some indoor spaces was brought into force on 30 November. Similar laws are in place in the devolved nations. para 4.

2 See, for example, Martin Z. Bazant and John W.M. Bush, 'A guideline to limit indoor airborne transmission of COVID-19', *Proceedings of the National Academy of Sciences*, 118.17 (2021). www.pnas.org/content/118/17/e2018995118.

3 World Health Organization, 'Archived: WHO Timeline'. www.who.int/news/item/27-04-2020-who-timeline---covid-19.

4 Dyani Lewis, 'COVID-19 rarely spreads through surfaces. So why are we still deep cleaning?', *Nature*, 29 January 2021. www.nature.com/articles/d41586-021-00251-4 para 7.

5 Derek Thompson, 'Hygiene theater is a huge waste of time', *The Atlantic*, 27 July 2020. www.theatlantic.com/ideas/archive/2020/07/scourge-hygiene-theater/614599/ para 7.

6 See, for example, Liz Lightfoot, 'Schools in England say government not providing enough air purifiers', *The Guardian*, 14 January 2022. www.theguardian.com/education/2022/jan/14/schools-england-demand-air-purifiers-being-underestimated.

7 Horwell, 'Ministers know which masks provide the best Covid protection—why not tell the UK public?', paras 1–2.

8 Horwell, 'Ministers know which masks provide the best Covid protection—why not tell the UK public?' para 8.

9 Katherine A. Craik, 'Sorting pistol's blood social class and the circulation of character in Shakespeare's 2 *Henry IV* and *Henry V*', in *Blood Matters*, edited by Bonnie Lander Johnson and Eleanor Decamp (Philadelphia, PA: University of Pennsylvania Press, 2018), pp. 43–58, p. 48.

10 Dympna Callaghan, '(Un)natural loving: swine, pets and flowers in *Venus and Adonis*', in *Textures of Renaissance Knowledge*, edited by Philippa Berry and Margaret Tudeau-Clayton (Manchester: Manchester University Press, 2003), pp. 58–78, p. 62.

11 Callaghan, '(Un)natural loving: swine, pets and flowers in *Venus and Adonis*', p. 64.

12 The OED records the first figurative use of the adjective meaning 'salacious; lustful, sexy, torrid' in *The Daily Telegraph*, 10 June 1970.

13 See, for example, William Keach, *Elizabethan Erotic Narratives: Irony and Pathos in the Ovidian Poetry of Shakespeare, Marlowe and Their Contemporaries* (New Brunswick, NJ: Rutgers University Press, 1976).

14 Stephen Guy-Bray, *Shakespeare and Queer Representation* (London: Routledge, 2021), p. 149.

15 For a compelling essay which 'argues that Adonis, and other eroticized adolescent male characters like him in early modern literature, might best be described as actively asexual rather than passively presexual, and discussed in terms of their successfully articulated disinterest in

sex and romance', see Simone Chess, 'Asexuality, Queer Chastity, and Adolescence in Early Modern England', in *Queering Childhood in Early Modern England*, edited by Jennifer Higginbotham and Mark Albert Johnston (London: Palgrave Macmillan, 2018), pp. 31–55, p. 31.

16 Wayne A. Rebhorn, 'Mother Venus: temptation in *Venus and Adonis*', *Shakespeare Studies* 11 (1978): 1–19, p. 2, p. 4. For an extended discussion of how Edmund Spenser's *The Faerie Queene* anticipates vampiric discourse, see Garrett Sullivan, 'Vampirism in the Bower of Bliss', in *Gothic Renaissance: A Reassessment*, edited by Elisabeth Bronfen and Beate Neumiere (Manchester: Manchester University Press, 2014), pp. 167–79.

17 Geoffrey Lloyd, 'Pneuma between body and soul', *The Journal of the Royal Anthropological Institute*, 13 (2007): 135–46, p. 137.

18 Samuel Taylor Coleridge, *Biographia Literaria* [1817] (New Zealand: Auckland: The Floating Press 2009), p. 251.

19 Abraham Fraunce, *The Third Part of the Countess of Pembroke's Ivychurch* (London, 1592), 45v. I was reminded of Fraunce's text in Sarah Carter, '"With kissing him I should have killed him first"': Death in Ovid and Shakespeare's *Venus and Adonis*", *Early Modern Literary Studies*, Special Issue 24 (2015): 1–13. p. 3. https://extra.shu.ac.uk/emls/journal/index.php/emls/article/view/274.

20 Randall Martin, *Shakespeare and Ecology* (Oxford: Oxford University Press, 2015), p. 29.

21 *The Norton Shakespeare*, p. 635.

22 Katherine Duncan-Jones and Henry Woudhuysen (eds), *Shakespeare's Poems* (London: Thomson Learning, 2007), p. 63.

23 Mary Floyd Wilson, *English Ethnicity and Race in Early Modern Drama* (Cambridge: Cambridge University Press, 2003), p. 2.

24 Mary Floyd-Wilson and Garrett A. Sullivan, Jr, 'Introduction: inhabiting the body, inhabiting the world', in *Environment and Embodiment in Early Modern England*, edited by Mary Floyd-Wilson and Garrett A. Sullivan, Jr. (Basingstoke: Palgrave Macmillan, 2007), pp. 1–13, p. 5.

25 Mary Floyd-Wilson, 'English mettle', in *Reading the Early Modern Passions: Essays in the Cultural History of Emotion*, edited by Gail Kern Paster, Katherine Rowe and Mary Floyd-Wilson, pp. 130–46, 133–6.

26 I'm referring here to Floyd-Wilson and Sullivan's term 'somatic ecology'. Floyd-Wilson and Sullivan, Jr, 'Introduction: inhabiting the body, inhabiting the world', p. 3.
27 Floyd-Wilson and Sullivan, Jr, 'Introduction: inhabiting the body, inhabiting the world', p. 5.
28 *The Norton Shakespeare*, p. 637n.
29 *The Norton Shakespeare*, p. 647n9.
30 OED 2a.
31 OED 7a. The OED notes that the earliest use of 'reek' in this context is from 1609. However, the first complete English version of one of Shakespeare's favourite sources, Arthur Golding's *The 15 Bookes of P. Ovidius Naso* (1567), uses 'reek' as part of its translation of Book 7's account of the plague at Aegina: 'first the Aire with foggie stinking reeke/ Did daily overdreepe the earth' (p. 89r).
32 Vivian Nutton, 'How new diseases were investigated in the 16th century', *The British Academy Blog*, 4 March 2021 www.thebritishacademy.ac.uk/blog/how-new-diseases-were-investigated-in-the-16th-century/ para 5.
33 Tim Ingold, 'Earth, sky, wind, and weather', *The Journal of the Royal Anthropological Institute*, 13 (2007): 19–38, p. 20, p. 33.
34 Nicola Davis, 'UK's autumn Covid wave could be worse than the last as cases rise', *The Guardian*, 23 September 2022. www.theguardian.com/world/2022/sep/23/uks-autumn-covid-wave-could-be-worse-than-the-last-as-cases-rise para 1.

Soil

Down to earth

Five

A few years before the combined pressures of Covid-19 and climate breakdown constrained carefree approaches to international travel, in June 2017 I journeyed with colleagues from Lancaster University's Environment Centre (LEC) to the University of Lausanne to discuss their collaborative course on global warming and societal change. As we travelled by road, air and rail from the north-west of England to the Swiss city on the shores of Lake Geneva, my LEC co-workers' expertise in soil science and my interest in Shakespearean surfaces led to interdisciplinary chats about this enigmatic brew of organic matter, minerals, gases, liquids and organisms. Those conversations have stayed with me and motivate much of my thinking in this section on Shakespeare, surfaces and soil. Specifically, Jessica Davies' research on 'sustainable soils, land and food systems across natural, agricultural, and urban landscapes' has encouraged me to take notice of this understated, underfoot habitat and helped me throw a spotlight on Shakespeare's treatment of this complex substance.[1] Peter Wohlleben explains how:

> For us humans, soil is more obscure than water, both literally and metaphorically. 'Whereas it is generally accepted that we know less about the ocean floor than we know about the surface of the moon, we know even less

DOI: 10.4324/9780429326752-6

about life in the soil.' ('Do trees have feelings too? One expert says they do—*The Telegraph*') Sure, there's a wealth of species and facts that have been discovered and that we can read about. But we know only a tiny fraction of what there is to know about the complex life that busies itself under our feet.[2]

Love's Labour's Lost's Holofernes, a figure generally depicted as a comic pedant fashioned to poke fun at premodern humanist education, hints at a specialised knowledge of the globe's manifold composition when he tells us that 'the face of *terra*'— a phrase which reaches for but doesn't quite get to the word 'surface'—means 'the soil, the land, the earth' (4.2.6). And Henry IV in *Henry IV Part One* notices that 'Sir Walter Blunt' is 'Stained with the variation of each soil' (1.1.63–5) between Holmedon in Northumberland and London. As a flat-dweller who has valiantly tried to keep one houseplant alive during lockdowns and can't remember when they last touched an earth-like substance out-of-doors, Wohlleben's comments rather than Holofernes' or Henry IV's resonate with my own impoverished hands-on acquaintance with soil. But I've come to realise that Holofernes' vernacular understanding of the Latin word *terra* [earth] alongside Henry IV's keen observation of soil's visual diversity form part of an essential ecological perspective.

Far from being the cladding out of which materials of obvious value are extracted—fossil fuels, precious metals, gemstones, pigments, root vegetables—I've discovered soil's inestimable importance:

> Globally, the soil contains over 3,000 gigatonnes of carbon, about four times the amount of carbon in the atmosphere

and the world combined. This vast underground store regulates the global carbon cycle, while contributing to food production, biodiversity, drought and flood resilience, and ecosystem functioning.[3]

Often thought of as a 'dull mass of ground-up rock and dead plants', an inanimate substance of mostly brown, sometimes soggy/ sometimes dusty matter, soil turns out to be 'fractally scaled…Bacteria, fungi, plants and soil animals, working unconsciously together, build an immeasurably intricate, endlessly ramifying architecture, that…organises itself spontaneously into coherent worlds'.[4] By the time this biological wonder stuff becomes dirt, it's dead and well on the way to becoming dust.[5] Before that terminal state, however, the integrated structure of soil's rhizosphere—a Deleuzean/ Guattarian dream made manifest—is crucial for resisting the obvious effects of climate collapse.

Alongside recent work by researchers and activists in environmental studies, essay collections such as Hillary Eklund's *Ground-Work: English Renaissance Literature and Soil Science* ('the first collection dedicated to the representations of soil in literature')[6] have developed my views on Shakespeare's soil. Randall Martin sees the 'agricultural details and fraternal rivalry' of *As You Like It*'s opening scene as representations of 'early modern capitalism's shift toward maximising crop yields and revenue from arable land', which made 'soil improvement and sustainability contentious issues'.[7] With this social allegory in mind, the fact that the two brothers are joined by their deceased father's servant Adam, Eden's gardener, is noteworthy. And it's evident that this late-Elizabethan comedy highlights soil's significance in less overtly patriarchal and combative ways too. Economic tensions related to but different from the ones identified by Martin crop up just a few scenes later when *As You Like It*'s aristocratic Celia transforms herself into the lower-born woman Aliena by dressing 'in poor and mean attire' and smearing her face 'with a kind of umber' (1.3.105–6). Celia's swift use of a substance usually formed from heated brown clay, which produces a 'havering hue' ranging from 'the deep redness of blood and the blasé blandness of mud'[8] embeds a speck of material history concerned with earth, extraction, aesthetics, and class into Shakespeare's Elizabethan comedy. By comparison, and as I'll discuss for the remainder of this section, the more meditative textures of three tragedies—*Richard II*, *Hamlet* and *Timon of Athens*—work around, think through and work against soil in respectively extended ways.

In a rich discussion of how *Richard II* treats the soil as a corporeal form 'governed by the humoral, biological, and seasonal movements of blood, tears, digestion, and spirit [which] needs to be understood as part of the play's broader concern with the human desire to regulate and control the turbulence of nature'.[9] Bonnie Lander Johnson's ecological lens draws attention to the play's bookish approach. Alongside the Bible, classical literature, chronicles and works about husbandry, Lander Johnson primarily shows how Elizabethan almanacs, 'guides on when to do those activities that formed the basis of early modern life in the best and most timely fashion',[10] are significant but hitherto neglected intertexts for *Richard II*. In many ways, then, the play could be seen as taking an Elizabethan Protestant humanist approach to soil, that is one that sees a vernacular Christian theology primary driven by words (scripture). If Lander Johnson speaks to 'soil's textuality'.[11] *Richard II* shows us how soil's materiality is subordinate to the period's orthodox aristocratic ideals.

Shakespeare's *Richard II* starts by showing the King's approach to civil unrest at a micro level and how that approach has a macro-cultural impact. The one-on-one dispute between Mowbray and Bolingbroke about the circumstances of the Duke of Gloucester's death results in exile for both, the acquisition of Bolingbroke's land and hoped-for 'coffers' (1.4.42) to help fund Richard II's wars against the Irish. John of Gaunt's prophetic set piece about England as a 'sceptred isle' in 2.1.40– 68 foregrounds the play's rhetorical and ritualistic treatment of landscape. Though powerful, John of Gaunt's lines betray a distant relationship to the actual fabric of England's earth that will reverberate through the play as whole. As a thinly veiled allegory about King Richard II's profligate treatment of Bolingbroke (Gaunt's son) and his kingdom, Gaunt doesn't actually speak about the nation's geological foundations in material terms. To be sure, Gaunt's famous speech isn't even geographically accurate: England obviously isn't an island. Though Stuart Elden's book *Shakespearean Territories* convincingly discusses how 'his plays, and some of his poetry, exhibit a profound geographical imagination' and that *Richard II* is especially interested in questions of law, economics, agriculture and exile,[12] it's worth keeping in mind that 'the subsumption of Wales and Scotland into England and the misrecognition of England as an island [...] has bedevilled imaginings of the geography and polity of these islands for centuries'.[13] Even if Gaunt's outlook for Richard's kingdom is different than those of his sovereign's, Richard II shares his uncle's textual treatment of soil, an underfoot substance which is both part of—and yet this play insinuates—largely divorced from ideas of landscape, earth and environment.

When he returns from Ireland and arrives at Edward I's former coastal fortress of Harlech Castle on the Welsh coast, part of the system of linked strategically placed castles commonly known as the 'iron Ring' and built to control Wales, Richard 'touches the ground' (3.2.5.s.d.) and addresses it directly. 'Dear earth,' he says, 'I do salute thee with my hand, / Though rebels wound thee with their horses' hoofs' (3.2.6). Over the next 17 lines, Richard embraces the land in a performance of regal possession first calling the ground 'my earth' and then issuing the instruction 'Feed not thy sovereign's foe, my gentle earth' (3.2.10, 12). He says that he 'weep[s] for joy' (3.2.4). Nonetheless, Richard's actions are part of a public statement rather than an impulsive outpouring of emotion. Once he's finished his oration to the earth, Richard turns to the attendant Duke of Aumerle and the Bishop of Carlisle saying 'Mock not my senseless conjuration, lords' (3.3.23). Indeed, Richard's speech to his inhuman and human auditors is part of a longer political broadcast upholding his divine right as 'an anointed king' (3.2.51) in what he thinks is a Welsh-supported comeback. But just a few lines later, Richard learns from the newly arrived Earl of Salisbury that 'all the Welshman...are gone to Bolingbroke, dispersed, and fled' (3.2.69–70). This information is swiftly followed by Sir Stephen Scrope's entrance, which brings more 'tidings of calamity' (3.2.101): 'both young and old rebel' (3.2.115) and the courtiers Bushy, Green and the Earl of Wiltshire have been executed. Richard turns to ponder his demise:

> Let's talk of graves, of worms and epitaphs,
> Make dust our paper, and with rainy eyes
> Write sorrow on the bosom of the earth.
> Let's choose executors and talk of wills—
> And yet not so, for what can we bequeath
> Save our disposèd bodies to the ground?

> Our lands, our lives, and all are Bolingbroke's;
> And nothing can we call our own but death,
> And that small model of the barren earth
> Which serves as paste and cover to our bones. (3.2.141–50)

In the first of his two third-act speeches about graves and 'the death of kings' (3.2.152), the scene shifts from Richard's opening interest in the ground to subterranean levels. Here, Richard takes control of dusty matter; the 'bosom of the earth' becomes a place to inscribe his mortal body's 'sorrow'. But even with that body interred within the soil, Richard's sovereignty aims to rise above it.

Act 3 Scene 3 begins outside Flint Castle about 60 miles from Harlech in north-east Wales. The men who will unseat the monarch, Bolingbroke, the Duke of York, the Earl of Northumberland plus soldiers 'with drum and colours' (3.3.s.d.) are assembled onstage. Bolingbroke asks Northumberland to deliver a proposition to the by now tenuously titled sovereign:

> Henry Bolingbroke
> Upon his knees doth kiss King Richard's hand,
> And sends allegiance and true faith of heart
> To his most royal person, hither come
> Even at his feet, to lay my arms and power,
> Provided that my banishment repealed
> And lands restored again, be freely granted.
> If not, I'll use the advantage of my power,
> And lay the summer's dust with showers of blood
> Rained from the wounds of slaughtered Englishmen;
> The which how far off from the mind of Bolingbroke

> It is such crimson tempest should bedrench
> The fresh green lap of fair King Richard's land,
> My stooping duty tenderly shall show. (3.3.34–47)

From the King's management of a range of disputes, from court (1.1.1) to Harlech, *Richard II* is shaped by monarchical ritual, ceremony and custom. Like John of Gaunt's and King Richard's earlier speeches, Bolingbroke abides by modes of early modern poetics which demand that sovereign power is enmeshed with the landscape, the environment and its inhabitants (a staple feature of the so-called body politic).[14] On the brink of exile at the end of 1.3, and as he bids 'farewell' to 'England's ground' and 'adieu' to the 'sweet soil', Bolingbroke speaks to the body politic that maintains his status as 'a true born Englishman' (1.3.269–72). In the militaristic space of Flint Castle, Bolingbroke's communication isn't constructed like John of Gaunt's, King Richard's or even his own earlier style of a florid speech. He makes plain how 'showers of blood/ Rained from the wounds of slaughtered Englishmen' will smother 'the summer's dust' and transform the 'fresh green lap of' markedly 'King Richard's land'. If Macbeth fears that his violent act of regicide will turn 'multitudinous seas incarnadine,/ making the green one red' (2.2.60), Bolingbroke aims to alter England's chromatic, cultural and climatic environment.

The scene quickly turns from the front of Flint Castle to Richard standing on the castle's walls. In between his open-air parley with Northumberland, and as he asks 'What must the King do now?' (3.3.142), Richard's more intimate dialogue with Aumerle returns to usurpation and their subsequent death. He will trade his 'large kingdom for a little grave':

> A little, little grave, an obscure grave;
> Or I'll be buried in the King's highway,
> Some way of common trade where subjects' feet
> May hourly trample on their sovereign's head,
> For on my heart they tread now, whilst I live,
> And buried once, why not up on my head?
> Aumerle, thou weep'st, my tender-hearted cousin.
> We'll make foul weather with despisèd tears.
> Our sighs and they shall lodge the summer corn,
> And make a dearth in this revolting land.
> Or shall we play the wantons with our woes,
> And make some pretty match with shedding tears;
> And thus to drop them still upon one place
> Until they have fretted us a pair of graves
> Within the earth. And therein laid? 'There lies
> Two kinsmen digged their graves with weeping eyes'.
> (3.3.152–68)

While Richard muses on his fatal outcome, he first considers his grave's diminishing size and then its possible unconsecrated location, the image of a self-fashioned earthly tomb produced 'with shedding tears' maintains a hoped-for sovereignty over soil itself. Markedly, no manual labour is envisioned here; the graves are simply 'fretted' via 'weeping eyes'. This funereal episode shows once more how the play's deposed King prefers rhetorical over material effect, a disposition underscored by the Gardener's aim in the very next scene to 'root away/ The noisome weeds which without profit suck/ The soil's fertility from wholesome flowers' (3.4.38–40). Whereas Richard's lyrical view of gravedigging avoids actual spade work, the Gardener's matter-of-fact style foregrounds a firsthand approach to soil management which *Hamlet* eventually magnifies.

True to his status as a student at Wittenburg University, Prince Hamlet asks a lot of questions. In fact, *he* suggests that he poses one of the most important questions of all: 'To be or not to be? That *is* the question' (3.1.58) (my emphasis). Hamlet's famous query is just one of many asked in this most interrogative play (which starts with the line 'Who's there?' (1.1.1)). Before reaching Act 3's existential crisis, Hamlet asks 'what is this quintessence of dust?' (2.2.298). While he doesn't actually ask 'what is soil?', the quotation suggests that 'we might see Hamlet as an early soil scientist… in relation to the mysterious natural processes of soil'.[15] Beyond character analysis, the play's structure moves from Prince Hamlet's propensity to think about the ground beneath our bodies to Act 5's dramatisation of bodies in the ground.

At the start of *Hamlet*'s final act, like Celia's use of umber to disguise her upper-class credentials,[16] the play shifts from the aristocratic worldviews aligned with the foregoing scenes in and around Elsinore Castle to those of 'two Clowns [rustics] carrying a spade and a pickaxe' (5.1 s.d.), soil's props. If Prince Hamlet thinks about soil, then these men clearly work with it. Their conversation is aware of social difference and their own genealogy forged via biblical doctrine rather than political theology and theories of sovereignty. 'There is no ancient gentleman', says First Clown, 'but gardeners, ditchers, and gravemakers; they hold up Adam's profession' (5.1.28–29). The 50 lines or so of dialogue between First and Second Clown establish these men's confidence in their place in early modern England's social hierarchy, a perspective at odds with Denmark's dislocated heir. More than that, First

Clown describes a community of workers linked by three soil-based occupations. It's striking, then, that in conversation with Hamlet, First Clown refers to himself as a 'sexton' (5.1.149), a job encompassing the role of gravedigger along with the wider remit of church management and maintenance in general. In so doing, from his reference to the biblical Adam through to his self-definition as a sexton, First Clown connects soil to both spiritual and societal realms in pragmatic rather than philosophical ways. The song accompanying his physical exertions is a good example of his practical demeanour. It begins:

> But age with his stealing steps
> Hath caught me in his clutch,
> And hath shipped me intil the land,
> As if I had never been such.
> [*He throws up a skull*] (5.1.66–69s.d.)

With these balladic lines, faintly inscribed with the familiar axiom from the Book of Common Prayer's burial service 'earth to earth, ashes to ashes, dust to dust', First Clown delivers a pop version of Prince Hamlet's philosophical quandaries about life, death and all the spaces in between.

Indeed, this Shakespearean figure—one that's literally 'down to earth'— functions as a foil to the intellectually driven Prince who has just returned to Denmark from exile in England. When First Clown's graveside singing offends Hamlet, for instance, Horatio remarks that 'Custom hath made it in him a property of easiness' (5.1.63), a description we'd be hard-pressed to apply to the play's grieving, vengeful and increasingly class-conscious titular protagonist. Hamlet reckons that 'hand[s] of little employment hath the daintier sense' (5.1.64–5) and the First Clown's ditty (measured by upthrown skulls) encourages a series of questions about the lives the dead once led, the last one addressed to the gravedigger himself: 'Whose grave's this, sirrah?' (5.1.107–8). A faltering conversation between the apparently 'daintier' Prince and the 'absolute' (5.1.126) worker leads to a soil-specific enquiry. 'How long', asks Hamlet, 'will a man lie i'th' earth ere he rot?' (5.1.151). First Clown answers:

> I' faith, if a be not rotten before a die—as we have many pocky corpses nowadays, that will scarce hold the laying in—a will last you some eight year or nine year. A tanner will last you nine year. (5.1.152–55)

The gravedigger's knowledge of how cadavers' post-burial degradation are diversely affected by the 'pocky corpses' produced by venereal disease or trades of tanning leather draws attention to early modern England's material disquiet about human interment. Shakespeare's contemporary Christopher Marlowe vividly addresses a related topic in *The Massacre at Paris*' (1592) treatment of Admiral Coligny, the Huguenot leader assassinated by Catholic adversaries in the religious wars of 1572. In scene 11, the stage directions read 'Enter two [soldiers] with the Admiral's body' and a discussion ensues about how best to dispose of the corpse. First, they consider 'burn[ing] him for an heretic', but this isn't an acceptable solution as 'his body will infect the fire, and the fire the air, and so [they] will be poisoned with him' (3–5).[17] The two soldiers then decide they can't 'throw him into the river' because 'twill' corrupt the water, and the water the fish, and by the fish [themselves] when [they] eat them' (8–6). One of them then suggests that they 'throw him into the ditch', but the other quickly demands that they 'hang him upon this tree' (10–12). The stage directions tell us that 'They hang him. Enter the Duke of Guise, Catherine the Queen Mother, and the Cardinal [with Attendants]'. The French aristocrats are generally pleased with the Admiral's suspended body. He 'becomes the place so well' that the Queen Mother 'could long ere this have wisht him there' (15–16). But there's a problem: 'th' airs not very sweet' (17). In the end, the Duke orders 'Sirs, take him away and throw him in some ditch' and they 'Carry away the dead body' (19). In this episode's elemental focus on the difficulties of dispensing with the effects of a rotting human—there are obvious concerns about infection transmitted by fire, water and air—the play poses a tacit query about the consequences for public health of leaving a corpse in an earth-bound ditch. Both *The Massacre at Paris* and *Hamlet* are animated by the zeitgeist of post-Reformation England. However, from our historical distance, these two dramatisations of death and decay are also borne out of material concerns such as a European epoch shifting from humoral to Cartesian thought. Hamlet's graveside's stance, posing questions that simultaneously acknowledge and push back at the dissolution of human identity in soil, hints at the anthropogenic ecological crisis to come.

By comparison with *Richard II*'s and *Hamlet*'s respective interweavings of soily strands into their plots, *Timon of Athens* is based on blunt extremes. Matthew Salisbury's review of the Royal Shakespeare Company's (RSC) 2018 production sums up the play's narrative arc well: 'It starts in gold and ends in soil'.[18] In its portrayal of the main character's descent from luxury to penury, *Timon of Athens* is alert to the co-dependent relationship between a community and its organic foundations. While Shakespeare's play is based on classical Timonian narratives such as those found in Plutarch's *Lives* and Lucian's *Dialogues*, the tragedy zooms in on the birth, life and death of a cynical recluse.

Timon of Athens' opening scenes depict a generous, some might say profligate, Timon who enjoys distributing the finer things in life to his friends. In turn, Timon's friends bestow lavish gifts upon him. Despite his steward Flavius' attempts to manage the household's accounts, Timon find himself in debt. But when Timon can't pay his numerous creditors and he's forced to ask his friends for help, no one comes forward. As he turns away from his Athenian associates, Timon stages one final feast with a twist: he serves steam and stones to show his 'knot of mouth-friends' (3.7.81) exactly what he thinks of them. Timon's inedible meal is accompanied by the verbal and physical abuse of his erstwhile companions and the evening ends with the host's powerful exiting couplet: 'Burn house! Sink Athens! Henceforth hated be/ Of Timon man and all humanity! (3.7.96–7). While the still-indifferent dinner guests ask 'Know you the quality of Lord Timon's fury?' (3.7.98), Act 4 stages the final stage of his transformation into 'Misanthropos' (4.3.53). On the other side of the city's walls, Timon's 41-line soliloquy curses Athens and its inhabitants:

> Itches, blains,
> Sow all th'Athenian bosoms, and their crop
> Be general leprosy! Breath infect breath,
> That their society, as their friendship, may
> Be merely poison!
> [*He tears off his clothing*]
>   Nothing I'll bear from thee

> But nakedness, thou detestable town;
> Take thou that too, with multiplying bans.
> Timon will to the woods, where he shall find
> Th'unkindest beast more kinder than mankind. (4.1.28–36)

Timon's broad awareness of humoralism's relationality lets him imagine respiration as a kind of early modern biological warfare: 'breath infect breath'. What he doesn't realise is that he can't divorce himself from the finely balanced environmental network he inhabits. That Timon's ambition is to end up in a grave within the 'beached verge of the salt flood' (5.2.101) instead of soil ultimately underscores this lack of humoral understanding and ecological nous.

We've already seen in *Hamlet* that the spade functions as soil's prop. This tragedy's penultimate act stages Timon's loam-laden interactions as he tries to find sustenance:

> *Enter Timon [from his cave] in the woods [half naked and*
>   *with a spade]*
> O blessèd breeding sun, draw from the earth
> Rotten humidity; below thy sister's orb
> Infect the air.
>   ...
>   Earth, yield me roots.
>   [*He digs*]
> Who seeks for better of thee, sauce his palate
> With thy most operant poison.
>   [*He finds gold*]
>   ...
>   Come, damnèd earth,
> Thou common whore of mankind, that puts odds

Among the rout of nations; I will make thee
Do thy right nature. (4.3.1–44)

It's clear that Timon's metamorphosis goes beyond misanthropy. In his hoped-for reclusive state, Timon doesn't need gold: he needs 'roots'. The soil doesn't provide what Timon now needs to live. And where once he denounced his fairweather friends, he now protests against the earth. In a discussion focused on food Simon C. Estok writes that:

> Timon certainly is a person of the Renaissance and sees the world as belonging to himself for his own consumption: "myself, who had the world as my confectionary" (4.3.259–60), have lost everything, he complains. He ought to have thought about his unsustainable lifestyle earlier: in this, he seems almost an allegorical critique of contemporary humanity and its conspicuous consumption. Relying on a framework that presumes exemption of the natural world from ethical consideration (an exemption connected with ecophobia), early modern English food sources…were noticeably transforming from the local to the international, a global confectionary.[19]

Estok's nuanced essay reads the play in twenty-first century hindsight, that is *Timon of Athens* addresses 'questions raised by the global-sourcing/local-demand conflict of Shakespeare's food through performance'.[20] For me, thinking about *Timon of Athens* in performance now takes a step further in addressing sustainable food production. Anticipating the *New York Time's* view that 'From stages to runways, dirt has become a go-to metaphor'[21] by a few years, Simon Godwin's soil-strewn 2018 production for the RSC suggests that *Timon of Athens* has the propensity to makes soil an onstage character in its own right.

A month before the UK recorded its first case of Covid-19, in December 2020 the *Guardian* reported on the United Nation's State of Knowledge of Soil Biodiversity, which warned that

> the worsening state of soils is at least as important as the climate crisis and destruction of the natural world above ground. Crucially, it takes thousands of years for soils to form, meaning urgent protection and restoration of the soils that remain is needed. The scientists describe soils as like the skin of the living world, vital but thin and fragile, and easily damaged by intensive farming, forest destruction, pollution and global heating.[22]

In the three plays I've discussed in this section, soil is a key part of each tragic protagonists' onstage lifespan. If *The Comedy of Errors*, *Twelfth Night* and *The Tempest* can be yoked together as Shakespeare's Shipwreck Trilogy (as in the RSC's 2012 season), given the urgency of raising awareness about the 'skin of the living world', I propose that *Richard II*, *Hamlet* and *Timon of Athens* are now known as Shakespeare's Soilscape Tragedies.

NOTES

1 Professor Jessica Davies, Lancaster University. www.lancaster.ac.uk/lec/about-us/people/jessica-davies.
2 Peter Wohlleben, *The Hidden Life of Trees*, translated by Jane Billinghurst (London: William Collins, 2017), p. 85.
3 Jenifer L. Soong, 'Soil', in *The Climate Book*, created by Greta Thunberg (London: Allen Lane, 2022), pp. 116–17, p. 116.
4 George Monbiot, 'The secret world beneath our feet is mind-blowing—and the key to our planet's future', *The Guardian*, 7 May 2022.

www.theguardian.com/environment/2022/may/07/secret-world-beneath-our-feet-mind-blowing-key-to-planets-future paras 6–7.

5 For a brilliant book-length study of this material, see Joseph A. Amatao, *Dust: A History of the Small and the Invisible* (Berkeley: University of California Press, 2000).

6 Hillary Eklund (ed), *Ground-Work: English Renaissance Literature and Soil Science* (Pittsburgh, PA: Duquesne University Press, 2017), p. 3.

7 Randall Martin, 'Fertility versus firepower', in *Ground-Work*, pp. 129–47, p. 130.

8 Kelly Grovier, 'Umber: The colour of debauchery', *BBC Culture*, 19 September 2018. www.bbc.com/culture/article/20180919-umber-the-colour-of-debauchery para 1.

9 Bonnie Lander Johnson, 'Visions of soil and body management: The Almanac in *Richard II*', in *Ground-Work*, pp. 59–78, pp. 59–60.

10 Johnson, 'Visions of soil and body management: The Almanac in *Richard II*', p. 70.

11 Johnson, 'Visions of soil and body management: The Almanac in *Richard II*', p. 70

12 Stuart Elden, *Shakespearean Territories* (Chicago, IL: University of Chicago Press, 2018), p. 1, p. 88.

13 Kate Chedgzoy, 'This pleasant and sceptered isle: insular fantasies of national identity in Anne Dowrich's *The French Historie* and William Shakespeare's *Richard II*', in *Archipelagic Identities: Literature and Identity in the Atlantic Archipelago, 1550–1800*, edited by Philip Schwyzer and Simon Mealor (Aldershot: Ashgate, 2004), pp. 25–42, p. 25–6.

14 The foundational book on the topic is Ernst H. Kantorowicz, *The King's Two Bodies: A Study in Medieval Political Theology* [1957], Princeton Classics Edition (Princeton, NJ: Princeton University Press, 2016).

15 Eklund, *Ground-Work*, p. 2.

16 The RSC's May 2013 repertory productions of *As You Like It* (directed by Maria Alberg) and *Hamlet* (directed by David Farr) are connected via soil: 'The same foundational level of muddy soil that was exposed throughout *Hamlet* to finish that play in an upturned graveyard emerged again here, but as the end result of the gradual unpacking of a formal, tautly controlled environment'. Peter Kirwan, '*As You Like It* (RSC) @ The Royal Shakespeare Theatre', *The Bardathon*, 5 May 2015. https://blogs.nottingham.ac.uk/bardathon/2013/05/05/as-you-like-it-rsc-the-royal-shakespeare-theatre/ para 1.

17 All quotations are from Frank Romany and Robert Lindsey (eds), 'The massacre at Paris', in *Christopher Marlowe: The Complete Plays* (London: Penguin, 2003), pp. 507–62.
18 Matthew Salisbury, 'Timon of Athens: Swan Theatre', *Stratford Observer*, 14 November 2018. https://stratfordobserver.co.uk/lifestyle/review-timon-of-athens-rsc-swan-theatre-stratford-upon-avon-10376/ para 2.
19 Simon C. Estok, 'Timon of Athens, food transformations, and the world as confectionary', *Kritika Kultura*, 33/34 (2019/2020), pp. 667–83, p. 672.
20 Estok, 'Timon of Athens, food transformations, and the world as confectionary', p. 669.
21 Nick Haramis, 'How did mud get everywhere? from stages to runways, dirt has become a go-to metaphor', *The New York Times*, 21 November 2022. www.nytimes.com/2022/11/21/t-magazine/mud-dirt-theater-fashion.html.
22 Damian Carrington, 'Global soils underpin life but future looks "bleak", warns UN report', *The Guardian*, 14 December 2020 https://www.theguardian.com/environment/2020/dec/04/global-soils-underpin-life-but-future-looks-bleak-warns-un-report.

Slime

Sensory plays

Six

Where there's soil there's often slime, 'soft glutinous mud; alluvial ooze; viscous matter deposited or collected on stones, etc.'[1] One of my earliest and most vivid memories takes place in the small, enclosed space behind my grandmother's two-bedroomed terraced house near Southampton docks. I'm about two years old and sprinkling water—with determined care—over the deep dusty edges of the concrete path that stretches the short distance from the backdoor of the house to the yard's end. The vignette closes with early years me joyfully squelching the resulting slime between my fingers.

In the six decades spanned by this reminiscence, slime in Anglo-American cultures has acquired a sometimes glamorous, sometimes troubling appearance alongside an increasing interest in its significance for human creativity. Originally produced by the toy company Mattel in 1976 and made from guar gum and borax, first-generation recreational slime was green (something akin to Pantone 368C) and sold in small plastic packages shaped like vintage dustbins. Twenty-first-century commercial slime, a substance generally made from polyvinyl alcohol and borate ion, has a wide-ranging colour palette and a vast array of finishes, fragrances and textures. Psychoanalytic criticism in the 1980s, by contrast, sees slime as abject, that Kristevan quality that 'confronts us, on the one

DOI: 10.4324/9780429326752-7

hand, with those fragile states where [the human] strays on the territories of animal…[and] on the other hand…with our earliest attempts to release the hold of maternal entity'.[2] Abject slime is thus associated with disquieting conditions, personal, political and planetary. On the latter, Simon C. Estok writes that:

> slime is becoming an inevitability in places where it should not be, where it is unnatural and just plain dangerous—a result beyond us that we cannot control for behaviors within us that we could not control.
>
> It is an open question whether or not UBC marine biologist Daniel Pauly is accurate to suggest instead of the term Anthropocene 'a new name of this new era, the age of slime' (Pauly 2010, p. 61). With the removal of the high end of food chains in global seas, jellyfish are proliferating, and oceans are becoming slime. Thus, Pauly proposes 'that [our era] be called "Myxocene"' (p. 61), the Age of Slime.[3]

Given Pauly's warning about the slimic state of the earth's oceans, climate emergency activists rightly question the production of commercial slime, essentially a plastic substance made for frivolous consumption.[4]

Other kinds of creative media provide further opportunities for thinking about slime's pejorative associations. Monomi Park's 2017 video game Slime Rancher, for example, features the titular character Beatrix LeBeau who gathers 'slimes' on Rainbow Island as players 'attempt to amass a great fortune in the business of slime ranching',[5] a description which calls to mind all manner of extractive economies and oppressive ideologies bound up with the word 'slime'. To help extend our critical thinking about social inequalities, Estok tells us how:

> Unmistakably gendered, raced, and classed, slime connects
> and disconnects us, to each other and to the world. Slime
> is political. Understanding how we understand and mediate
> it can help us to move forward along the road to ecological
> responsibility.[6]

Covid-19 amplifies slime's commodification and—importantly—its tendency to encourage critical and creative self-reflection about the earth and humankind's engagement with their environments. Paraphrasing Estok, this section suggests that 'understanding how [Shakespearean texts] understand and mediate [slime] can help us to move forward along the road to ecological responsibility'. Slime, I argue, incites dynamic problem-solving through playful explorations and studying plays written by Shakespeare.

At the start of the UK's first lockdown in March 2020, together with crafting materials in general, sales of slime kits doubled.[7] And slime's profit-making popularity doesn't rely on a hands-on approach to the product. About 18 months into the pandemic, *The Huffington Post* discussed how watching Tik Tok and YouTube videos of slime allows spectators to 'zone out to the sight of hands poking, squishing and pulling multi-color slime like it's taffy. The different sounds emitted by slimes, such as popping and clicking, can lull viewers into a relaxed state'.[8] Slime's recent appeal for leisure activities such as these led to discussions about its significance for well-being and Autonomous Sensory Meridian Response (ASMR).[9] Indeed, the current upswing in what's been called the 'slime economy'[10]—which includes the publication of Susanne Wedlich's *Das Boch vom Schleim* (2019), translated from German by Ayça Türkoğlu as *Slime: A Natural History* in 2021,[11] and theoretical interest in 'slime dynamics'[12]—indicates that popular

Western culture has only just begun to consider this enigmatic stuff's potential. As Wedlich/ Türkoğlu state, '[Slime] is protean in its behaviour; it is the material of interfaces and has a unique place in our imaginations'.[13] It turns out that twenty-first century Shakespeare Studies participates in these analytical and inventive discourses about slime too.

Physical scientists, specifically rheologists, have studied this ubiquitous non-Newtonian fluid at length. Like the other colloids discussed in this book (smoke, clouds, soil, steam, silk), slime's solid/fluid state confounds notions of surface stability in material and metaphoric terms. Anticipating Estok's and Wedlich/ Türkoğlu's work by half a decade, Dan Brayton's 2016 essay 'Shakespeare and Slime: Notes on the Anthropocene' is an important starting point for any discussion of the playwright's semiotics of slime. From Aristotle's theories of spontaneous generation[14] to Paul Crutzen's understanding of the Anthropocene, Brayton points out that 'slime occupies the conceptual space where the human imagination begins to grasp, tentatively and tenuously, the materiality of life itself'.[15] More specifically, Brayton argues that

> Shakespearean representations of slime and its cousin ooze can be seen to represent an 'index of obscure prologue' to modern ecological thought, in general, and ecocriticism in particular (*Othello* 2.1.205–6). In certain instances this murky substance forms a matrix for the historical emergence of (early) modern ecological thought; the parenthesis is meant to suggest that Shakespeare posed in the realm of imaginative literature similar questions about life, matter, region, and place that ecologists would pose as a matter of empirical study in the nineteenth, twentieth and twenty-first centuries.[16]

Shakespeare's works use the words 'slime' or 'slimy' in just four plays: *Titus Andronicus, Richard III, Othello* and *Antony and Cleopatra*. Brayton's 'Shakespeare and Slime' focuses on *Antony and Cleopatra* and *Richard III*. A text in which Egypt's river Nile is a locus of spontaneous generation, Brayton's comments on *Antony and Cleopatra*—from Antony's ultimately fatal slime-sworn oath to Cleopatra in Act 1 (3.69) to the First Gentleman's coolly forensic approach to 'an aspic's trail' of 'slime' in the closing scene's depiction of the Egyptian Queen's death (5.2.340–1)—show how Shakespeare's 'thematics of slime…bespeak an exotic, excremental dimension of matter that has the capacity to generate life'.[17] Though this seventeenth-century play engages with slime's fecundity, Brayton argues that 'elsewhere in the Shakespeare corpus slime is more squarely associated with death. This is particularly the case in *Richard III*…where slime coats the bottom of the sea along with emblems of human vanity and cupidity'.[18] Accordingly, Brayton's thrilling ecological analysis of the Duke of Clarence's briny nightmare and the episode's interwoven images of 'inestimable stones', 'the sea' and 'dead men's skulls', which culminate in the 'slimy bottom of the deep' (1.4.24–33), foregrounds slime's capacity to delineate 'the boundary between life and death and also that between water and ocean bottom'.[19] Whereas Brayton focuses on environmental matters from an Anthropocenic perspective, Sophie Chiari's equally valuable 2019 discussion of 'Clime and Slime in *Antony and Cleopatra*' brings classical and premodern contexts into focus.[20]

Chiari unpacks ideological tenets found in Aristotle's *Politiques* (tr. 1598), José de Acosta's *Natural and Moral History of the East and West Indies* (1590, tr. 1604) and Jean Bodin's *The Six Books of the Commonwealth* (1576, tr. 1606). In doing so, Chiari shows how *Antony and Cleopatra* unsettles geohumoral racialist thought, which insists that 'body and character are formed in response to the environment'.[21] Chiari's important work discusses the play's engagement with the Ptolemaic universe's concept of seven concentric spheres with the 'huge sphere' (2.7.13) of the earth at the centre. In turn, and as de Acosta believes, the earth is divided into five climatic zones:

> The temperature of the middle region of the world, where the sunne continually runnes his course, is scorched and burnt up as with a neere fire. Joyning to the same region, there are two others of eyther side, which (lying betwixt the heat of this burning zone & the cruell cold of the other two extreames,) are very temperate, and can have no communication one with another, by reason of the excessive heate of the heaven.[22]

Following classical precedents, sixteenth- and seventeenth-century thinkers ascribe characteristics according to climatic zones. For instance, some early moderns believe that the sun's heat causes inhabitants of the South 'to have commonly sharper wits, but smaller courage' while the inhabitants of the colder Northern climate 'are very courageous, but of small wit and prudence'.[23] I have more to say on this topic in section 'Nine: Skin'. For now, I want to focus on Chiari's perceptive understanding of how *Antony and Cleopatra*'s muddy images join rather than divide global selfhoods. In this Roman tragedy, Chiari argues:

> Shakespeare…blurs all binaries and his slimy, saturnalian Alexandria, indulging in spectacles and lavish entertainments, is manifestly closer than austere Rome to the pomp of early modern (and muddy) London.[24]

Examining the loamy surface's carousing characteristics for Alexandrians and Londoners alike, Chiari's sense of slime's aporetic function in *Antony and Cleopatra* (as in Brayton's earlier essay) leads hierarchical binary oppositions—for example 'Rome/Egypt, Antony/Cleopatra, west/east, war/peace, civic duty/personal pleasure'[25]—to fracture if not to fall. As we have already seen, slime thinkers usually mention muddy matters' ability to dissolve identities and foster creativity in some way. But what's so striking about Chiari's essay is its focus on slime's importance for textual production itself.

In her discussion of the Nile's capacity as a key site for spontaneous generation, Chiari mentions that it's 'a marvellous absurdity [which] seemed to have mainly been the stuff of dreams and poets'.[26] Chiari returns to slime's writerly dimensions in her observations about the Nile's alembic qualities. When Antony observes how 'the seedsman/ Upon the slime and ooze scatters his grain/ And shortly comes to harvest' (2.7.20–22), Chiari writes:

> the ooze of the Nile is neither debasing nor destructive, for it simultaneously shapes and dissolves things and beings alike, melting and changing them as swiftly as clouds in the air. Like the writer's ink, this mud is the material element which allows poetry to come to life. The 'seedsman' throwing his 'grains' in the mud is therefore no mere harvester; he is also an image of the poet playing with words and filling up the void/blank of the page.[27]

My thoughts on slime take up and develop Chiari's insightful slimic analysis of *Antony and Cleopatra*. If this play represents the apex of Shakespeare's contribution to 'being and slime'[28]—*Antony and Cleopatra* is the last Shakespearean text to mention the word—the remainder of this section considers two things: how Shakespeare's treatment of slime in *Richard III* and *Titus Andronicus* engage with what seems to be a sixteenth-century interest in slime and textual production; what these early plays might tell us about slime's significance for the creation of Shakespearean surfaces themselves.

Figuratively, there's no joy attached to slime in these two Tudor tragedies. In many ways, *Richard III* and *Titus Andronicus* engage with Thomas Nashe's contemporaneous view of slime in his meandering yet marvellous mess of ideas about *The Terrors of the Night* (1594). Linked broadly to the same ideas of spontaneous generation underpinning Antony's appraisal of the Nile's prolificacy, Nashe explains how intense fear is made manifest at night by way of earth and water, elemental 'spirits', which

> feeding foggy-brained melancholy, engender thereof many uncouth terrible monsters. Thus much observe by the way, that the grossest part of our blood is the melancholy humour, which in the spleen congealed, whose office is to disperse it, with his thick steaming fenny vapours casteth a mist over the spirits, and clean bemasketh the fantasy.
>
> And even as the slime and dirt in a standing puddle engender toads and frogs and many other unsightly creatures, so this slimy melancholy humour still still thickening as it stands, *still engendreth many mis-shapen objects in our imaginations*. (my emphasis)[29]

Whether viewed as 'a sceptical account of dreams and apparitions'[30] or an 'attack on demonology',[31] *The Terrors of the Night* participates in what has been called 'the golden age of melancholy', a complex human disposition that 'might simultaneously be about being sensitive, being a snob, being constipated, being political, and being funny'.[32] Sixteenth-century melancholy, as Nashe's text illustrates, is also about the creative aspects of conjuring night-time visions. In fact, *The Terrors of the Night* defines this specific sleep scenario as a site of spontaneous generation.

Towards the end of his wandering wonderings, Nashe returns to melancholy's consistency:

> In all points our brains are like the firmament, and exhale in every respect the like gross mistempered vapours and meteors, of the more feculent combustible airy matter whereof affrighting forms and monstrous images innumerable are *created*, but of the slimy unwieldier drossy part, dull melancholy or drowsiness. (my emphasis)[33]

As I explained in 'Three: Sky', Shakespeare's Hamlet likens his mood to the 'o'erhanging firmament', which seems to him like 'a foul and pestilent congregation of vapours' (2.2.291–3), a demeanour which evidently chimes with *The Terrors of the Night*'s thoughts on 'melancholy humour['s]… fenny vapours' and how 'In all points our brains are like the firmament'. If the Prince's disposition alongside his midnight encounter with King Hamlet's ghost in the opening act (1.5.1–91, 152–83) personifies a premodern melancholic's propensity for unsettling nocturnal sights,[34] Nashe's prose offers more information about melancholy's humoral texture itself; it's thick and slimy. And even if this night-time ooze is aligned with Kristevan abjection, Nashe's melancholic slime also encourages the formation of 'objects in our imaginations' and facilitates humanism's reach for the means to describe those objects. Slime enables creativity—at least for some.

That fifteenth- and sixteenth-century Europe's interests in such ideas are gendered is not surprising:

> Though women who suffered from the disease are more scarce than men in the medical literature, melancholy itself is a female figure.... The Latinised Greek word *melancholia* is a feminine noun, and melancholy's deep association with the contemplative, as opposed to the active, life connects it with Renaissance ideas of feminine behaviour. In Albrecht Dürer's 'Melancolia I' [1514] ..., it is a female winged figure who sits in a sad and meditative pose, surrounded by items symbolic of artistic skill, higher knowledge, time, death, and physical decay.[35]

Dürer's allegorical artwork thus offers a detached, androcentric portrait of melancholy's indolence. Taking a different tack, Nashe's later view of melancholy's sliminess speaks to an embodied condition bound up with rather than separate from creative processes. No matter that the sentence's thrice-uttered word 'still' insinuates increasing inertia, 'slimy melancholy humour still still thickening as it stands', Nashe writes, 'still engendreth many mis-shapen objects in our imaginations'. As a medium that brings the body evidently into view onstage and as plays contemporaneous with *The Terrors of the Night*, both *Richard III* and *Titus Andronicus* invoke slime in Nasheanlike ways that accentuates creative praxes—for men.

While Brayton draws attention to the Duke of Clarence's nightmare in terms of slime's 'capacity to generate life', my view of Shakespeare's slime marks how Clarence 'passed a miserable night,/ So full of fearful dreams, of ugly sights (1.4. 2–3), a description which chimes with Nashe in mode and mood. Framed for treason by Richard, Duke of Gloucester, the Duke's dream comes from an obviously troubled space. Indeed, when Sir Robert Brackenbury opens the scene with the question 'Why looks your grace so heavily today?' (1.4.1), he lets the audience know that the scene is one of melancholy.[36] Alongside *The Terrors of the Night*, Clarence's account of his fitful sleep becomes an example of Nashean terror:

> Methoughts I saw a thousand fearful wrecks
> Ten thousand men that fishes gnawed upon,
> Wedges of gold, great ouches, heaps of pearl,
> Inestimable stones, unvalued jewels,
> All scattered in the bottom of the sea.
> Some lay in dead men's skulls; and, in those holes
> Where eyes did once inhabit, there were crept—
> As 'twere in scorn of eyes—reflecting gems,
> Which wooed the slimy bottom of the deep
> And mocked the dead bones that lay scattered by. (1.4.24–33)

This unsettling watery nightmare, which 'engendreth...misshapen objects' like *The Terrors of the Night*, turns out to be prophetic: Act 1 Scene 4 ends with the Duke of Clarence's murder by stabbing and drowning in a vat of wine. Brayton's nuanced reading makes plain this episode's 'benthic imagination'.[37] I want to argue that the play itself comes out of these Shakespearean lines of slimy terror. To begin with, the Duke of Clarence's account of nightmare weaves text. Next, without the Duke of Clarence's death Richard, Duke of Gloucester, cannot become Richard III and the play cannot unfold.

By comparison with the slimy scenario that appears in *Richard III*'s opening Act, Shakespeare's reference to slime crops up in the middle of his most violent play that often seems like a nightmare writ large: *Titus Andronicus*. The stage is carefully set up for slime's appearance. Rome's ageing soldier Titus has just witnessed his sons Martius and Quintus 'bound, passing [over] the stage to the place of execution' (3.1.s.d.) and has failed in his strenuous plea before the Tribunes to take pity on his 'condemnèd sons/ Whose souls', he begs, are 'not corrupted as' 'Tis thought' (3.1.8–9). Thwarted in his efforts to save his only sons out of 'two-and-twenty' who didn't die in 'lofty honour's bed' (3.1.10–11), alone onstage Titus turns to address non-human forms:

> O earth, I will befriend thee more than with rain
> That shall distil from these two ancient ruins,
> Than youthful April shall with all his showers.
> In summer's drought I'll drop upon thee still.
> In winter with warm tears I'll melt the snow
> And keep eternal springtime on thy face,
> So thou refuse to drink my dear sons' blood. (3.1.16–22)

Linking human grief with environmental precarity, Titus vows to use his tears as ecosystemic support before turning, once again, to petition the Tribunes without success. Finding some brief solace in the immediate landscape rather than his fellow countrymen, Titus replies:

> Why, 'tis no matter man. If they did hear,
> They would not mark me; if they did mark,
> They would not pity me; yet plead I must.
> Therefore I tell my sorrows to the stones,
> Who, though they cannot answer my distress,
> Yet in some sort they are better than the Tribunes
> For that they will not intercept my tale.

> When I do weep they humbly at my feet
> Receive my tears and seem to weep with me,
> And were they but attirèd in grave weeds
> Rome could afford no tribunes like to these.
> A stone is soft as wax, tribunes more hard than stones.
> A stone is silent and offendeth not,
> And tribunes with their tongues doom men to death.
>
> (3.1.33–46)

In a play that values speech and its metonym the tongue—an organ so brutally severed from Titus's daughter Lavinia—lithic silence gives Titus an audience and a sense of narrative control. But it's slime, not stone, that pushes the story forward.

Moments later, Lucius tells his father that he is now banished from Rome for trying to rescue his brothers (3.1.48–50). And there is more misery to come. Just eight lines along, after finding her traumatised at 2.4.10 by the brutal rape and mutilation carried out by the young Goths Demetrius and Chiron, Titus' brother Marcus arrives with his niece Lavinia. Not content with sexual assault and understanding all too well their analogue to Tereus' monstrous, mythic crime against Philomela in Book 6 of the *Metamorphoses*, the Goths cut off Lavinia's hands and cut out her tongue so that she can't name her attackers manually or in an oral form. In Act 3, Lavinia's father, uncle and brother try to work out what happened to her. While Marcus delivers a few lines summarising Lavinia's horrific plight and Lucius tries asking his sister a couple of direct questions, Titus takes up and develops the passionate elemental terrain with which he began Act 3. 'He that wounded her', he says:

> Hath hurt me more than had he killed me dead;
> For now I stand as one upon a rock
> Evironed with a wilderness of sea,
> Who marks the waxing tide grow wave by wave,
> Expecting ever when some envious surge
> Will in his brinish bowels swallow him. (3.1.92–97)

Lavinia's abused body demands our focus. Yet in *Titus Andronicus'* central act her father's increasingly anguished state is shown via Titus' extravagant speech and his use of non-human imagery, which shifts from speaking with stones in 3.1.36 to standing on sea-surrounded rocks:

> Gentle Lavinia, let me kiss thy lips;
> Or make some sign how I may do thee ease.
> Shall thy good uncle, and thy brother Lucius,
> And thou, and I, sit round about some fountain,
> Looking all downwards to behold our cheeks
> How they are stain'd, as meadows, yet not dry
> With miry slime left on them by a flood?
> And in the fountain shall we gaze so long
> Till the fresh taste be taken from that clearness,
> And made a brine-pit with our bitter tears?
> Or shall we cut away our hands, like thine?
> Or shall we bite our tongues, and in dumb shows
> Pass the remainder of our hateful days? (3.1.120–32)

Unlike Clarence's pithy retelling of his terrible dream, Titus calls forth lurid scenes of social destruction. Whereas *Richard III* uses slime in its adjectival form to describe an earthy/watery surface, Titus' addition of the adjective 'miry' to 'slime' makes links with biblical anguish such as Psalm 40's meditation on despair, which speaks of transcending 'the horrible pit', leaving 'the mud and mire' and 'setting…feet upon the rock'.[38] There's more that could be said about slime and its cognate terms in the Christian Bible. I want to keep my focus, however, on secular slime.

As with the irruption of the word 'slimy' into *Richard III*'s opening Act, slime's Act 3 entrance in *Titus Andronicus* occurs at a pivotal moment for the play itself. In a speech that's punctuated by questions, Titus' speech ends with a further question followed by a resolution: 'What shall we do? Let us that have our tongues/ Plot some device of further misery, /To make us wondered at in time to come' (3.1.133–5). With a nod here to the Ovidian poem which pulses through *Titus Andronicus* (the *Metamorphoses* ends with the Latin poet's wish for his 'Life' to be 'everlastingly… lengthened still by fame'),[39] a key part of that 'wonder' is the production of the 'plot', the telling of the story at large and, importantly, its role in keeping the memory of the Andronici alive. Or, if you like, the repeated stagings of Shakespeare's *Titus Andronicus*.

If we follow the Shakespearean slime trail forward from the early 1590s through to *Othello* and *Antony and Cleopatra*, with hindsight we might see that its appearance in the four plays changes. From its entrance in *Richard III* 1.4 to *Titus Andronicus* 3.1, for example, 'slime' crops up in the last 20 minutes or so of *Othello*:

> 'twas [Iago] that told me [about Desdemona] first.
> An honest man he is, and hates the slime
> That sticks on filthy deeds. (5.2.154–56)

By connecting 'slime' with 'filthy deeds', the tragedy's catchy epigram reduces slime's complexity to a singular marker of abjection. Whereas *Richard III* and *Titus Andronicus* suggest that slime is a tacit indicator of creative ignition and *Antony and Cleopatra* dramatises a more fulsome exploration of slime's unsettling powers, slime's iteration in *Othello* 5.2 prefaces the death of a protagonist known as 'the great storyteller'[40] and, of course, the end of the play itself.

This section has traversed a range of oozy terrains, from happy childhood memories to Covid-19 to criticism to creative works. Over four plays, and anticipating contemporary culture's fascination with this sticky substance, Shakespearean slime asks its audience to mull over a word that embraces both creative and destructive meanings. Like all signs after Saussure,[41] slime takes on cultural freight according to its context. Nonetheless, Shakespeare's engagement with the supple meanings of this colloidal matter helps us to glimpse—just for a moment—the significance of slime, signs and semiotic play for creative thinking, unpacking structural inequalities and making texts themselves.

NOTES

1. OED, slime, n. 1a.
2. Julia Kristeva, *Powers of Horror: An Essay on Abjection*, translated by Leon S. Roudiez (Columbia: New York, 1982), p. 13.
3. Simon C. Estok, 'The slimic imagination and elemental eco-horror', *Zeitschrift für Anglistik und Amerikanistik*, 70.1 (2022): 59–74, 67.
4. 'Slime: Can it be environmentally friendly?', *CBBC Newsround*, 18 September 2018. www.bbc.co.uk/newsround/45560860 para 6.
5. Monomi Park Press Kit. http://press.monomipark.com/sheet.php?p=slime_rancher para 4.
6. Estok, 'The slimic imagination and elemental eco-horror', p. 70.

7 Sarah Butler, 'Sales of board games and jigsaws soar during coronavirus lockdown', *The Guardian*, 1 April 2020. www.theguardian.com/business/2020/apr/01/sales-of-board-games-and-jigsaws-soar-during-coronavirus-lockdown para 8.
8 Tessa Flores, 'Can playing with slime actually benefit your mental health?', *The Huffington Post*, 11 October 2021. www.huffingtonpost.co.uk/entry/slime-mental-health_l_61607c96e4b0cc44c50c70d1 para 3.
9 See, for example, Flores, 'Can playing with slime actually benefit your mental health?'
10 Ema Sagner, 'The rise of the slime economy', NPR, 1 October 2017. www.npr.org/2017/10/01/552422040/the-rise-of-the-slime-economy.
11 Susanne Wedlich, *Slime: A Natural History*, translated by Ayça Türkoğlu (London: Granta, 2021).
12 Ben Woodard, *Slime Dynamics: Generation, Mutacion, and the Creep of Life* (Winchester: Zero Books, 2012).
13 Wedlich, *Slime*, p. 2.
14 For a book-length study see Daryn Lehoux, *Creatures Born of Mud and Slime: The Wonder and Complexity of Spontaneous Generation* (Baltimore, MD: Johns Hopkins, 2017).
15 Dan Brayton, 'Shakespeare and slime: notes on the Anthropocene', in *Ecological Approaches to Early Modern Texts: A Field Guide to Reading and Teaching*, edited by Jennifer Munroe, Edward J. Geisweift and Lynne Bruckner (London: Routledge, 2016), pp. 81–90, p. 81.
16 Brayton, 'Shakespeare and slime: notes on the Anthropocene', p. 81.
17 Brayton, 'Shakespeare and slime: notes on the Anthropocene', p. 87.
18 Brayton, 'Shakespeare and slime: notes on the Anthropocene', p. 87.
19 Brayton, 'Shakespeare and slime: notes on the Anthropocene', p. 88.
20 Sophie Chiari, 'Clime and Slime in Antony and Cleopatra', *Shakespeare's Representation of Weather, Climate and Environment* (Edinburgh: Edinburgh University Press, 2019), pp. 176–216.
21 Michael Schoenfeldt, 'How gardens feel: the natural history of sensation on Spenser and Milton', in *The Geography of Embodiment in Early Modern England*, edited by Garrett Sullivan and Mary Floyd-Wilson (Oxford University Press, 2020), pp. 14–38, p. 21.

22 Cited in Chiari, *Shakespeare's Representation of Weather, Climate and Environment*, p. 178.
23 Cited in Chiari, *Shakespeare's Representation of Weather, Climate and Environment*, p. 178.
24 Chiari, *Shakespeare's Representation of Weather, Climate and Environment*, p. 204.
25 Brayton, 'Shakespeare and slime: notes on the Anthropocene', p. 85.
26 Chiari, *Shakespeare's Representation of Weather, Climate and Environment*, p. 188.
27 Chiari, *Shakespeare's Representation of Weather, Climate and Environment*, p. 191.
28 This phrase is indebted to Iain Hamilton Grant's Heideggerian-inspired essay 'Being and slime: the mathematics of protoplasm in Lorenz Oken's "physio-philosophy"', in *Collapse: Philosophical Research and Development Volume IV*, edited by Robin Mackay, revised edition (Cambridge, MA: MIT Press, 2012), pp. 287–321.
29 Thomas Nashe, *The Terrors of the Night* [1594], edited by Nina Green, *The Oxford Authorship Site* (2002). www.oxford-shakespeare.com/Nashe/Terrors_Night.pdf p. 9.
30 The British Library, 'Literature: Shakespeare and Renaissance'. www.bl.uk/collection-items/the-terrors-of-the-night-by-thomas-nash-1594 para 1.
31 The Editors of *Encyclopaedia Britannica*, 'Thomas Nashe', *Encyclopedia Britannica* (2023). www.britannica.com/biography/Thomas-Nashe para 5.
32 Erin Sullivan, 'Melancholy: humorous identity and the allure of genius', in *Beyond Melancholy: Sadness and Selfhood in Renaissance England* (Oxford: Oxford University Press, 2016), pp. 87–125, p. 88, p. 125.
33 Nashe, *The Terrors of the Night*, p. 25.
34 Erin Sullivan tells us that 'Though we now tend to associate the condition with *Hamlet* and its tragic Prince, it was Shakespeare's comedies that took up the trope of melancholy most regularly, and laughed at it as they did so'. Sullivan, 'Melancholy: humorous identity and the allure of genius', p. 110.
35 Mary Ann Lund, *A User's Guide to Melancholy* (Cambridge: Cambridge University Press, 2021), p. 113.
36 *The Norton Shakespeare* p. 565n glosses 'heavily' as 'melancholy'.

37 Brayton, 'Shakespeare and slime: notes on the Anthropocene', p. 87.
38 Geneva Bible (1599), 40: 1–2.
39 Arthur Golding (tr.), *Ovid's The Metamorphoses*, edited by Madeleine Forey (London: Penguin, 2002), 15, 994–5.
40 Monica Beckner Robinson, 'The power of words: Othello as storyteller', *Storytelling, Self, Society*, 7.1 (2011): 63–71, p. 63.
41 For an excellent introduction to Saussurean structuralism, see Jonathan Culler, *Ferdinand de Saussure*, revised edition (Ithaca, NY: Cornell University Press, 1991).

Snail

Finding our place

Seven

A debilitating flu-like illness forced Elisabeth Tova Bailey to relocate from her farmhouse and regular walks with her dog in the nearby forest to a studio apartment kitted out for her bedridden convalescence. During her time in this new living space, which felt like being 'trapped inside a stark white box',[1] Bailey discovered that a forest snail, a *Neohelix albolabris* to be precise, was the ideal companion for her own body 'felled by a mysterious viral or bacterial pathogen'.[2] In the acclaimed 2010 book about her extraordinary year with this unassuming gastropod, Bailey writes of how 'The tiny, intimate sound of the snail's eating gave [her] a distinct feeling of companionship and shared space' and she marvels at the creature's ability 'to defy physics…Each morning the terrarium glistened with the silvery trails of its nighttime travels'.[3] At the start of the UK's first lockdown, I found Bailey's book about interspecies cohabitation helpful for coping with the isolation of solo-flat dwelling alongside the low-level thrum of my pandemic-induced anxiety. There are immeasurable differences of course for the reasons behind her seclusion and the one I experienced: Bailey was recovering from infection while I was trying to avoid contracting Covid-19. Regardless of the grounds for our respective seclusions, what I found most valuable about her writing was the tempo of the

DOI: 10.4324/9780429326752-8

non-human/human relationship that developed as her book progressed. Like the snail itself, the narrative's purposeful slowness soothed this reader's unquiet thoughts about the viral organisms proliferating around the earth.

Bailey's tender depiction of the lustrous trace left on the container's glass surface belies the complexity of the snail's dried mucus, which is comprised of two kinds of fluid: one to help its reproduction cycle and another to aid its gravity-flouting progress. In the twenty-first century, snail secretion (often referred to as 'mucin') is valued for its ability to restore damaged human skin and reports of 'milking "happy" snails for beauty products'[4] circulate alongside ethical concerns for helicicultural practices.[5] But Bailey's fascination with the glistening surficial remains of her snail's nocturnal activities takes an obviously different turn than snail farming's human-centred business. Her careful contemplation of this common mollusc shares some qualities with the following quotation from Thomas Nashe's *The Unfortunate Traveller* (1594):

> What need the snaile care for eyes, when he feeles the waie with his two hornes, as well as if hee were as sharp sighted as a decypherer.[6]

There are differences. Biologists now know that snail tentacles (from the Latin tentare, 'to feel') should really be called 'rhinophore' (from the Greek words rhinos [nose] and phorus [bearer]).[7] And with reference to the preceding section's focus, please be aware: the sixteenth-century snails I discuss are not slimy. Indeed, these earlier observations focus on a very different aspect of the creature's anatomy than Bailey's above. Nonetheless, Nashe's tentacular text similarly ponders the snail's movements to think about other ways of being in the world, to consider an experiential rather than visual mode of existence.

To Bailey's and Nashe's brands of snail scrutiny (a far less dangerous activity than 'The Snail-Watcher' in Patricia Highsmith's short story of that name who fell foul of 'sealing, murderous snail bodies'[8]), we could add a pictorial image from one of the most influential premodern texts about travel in English. The busy titlepage of Samuel Purchas' *Hakluytus posthumus or Purchas his pilgrimes* (1625) includes a snail with a globe for a shell approaching the North Pole as it moves clockwise across the earth's surface. Illustrating the same ability as Bailey's and Nashe's respective snails to 'defy physics' and traverse the earth's curved sphere, this gastropod clings onto and crawls across a surface. This snail's extraordinary qualities are captured by the accompanying motto 'It peregre estque domi: Terras fert fertur ad Astra' [He is at home and is abroad; He carries the earth to the stars]. Transporting the nifty threshold space of an exoskeletal shelter fashioned as a mortal map, the snail itself occupies a liminal place between terrestrial and celestial globes. The small detail from Purchas' *Hakluytus Posthumous* captures both an Anglocentric curiosity for these creatures and their capacity to facilitate humankind's understanding 'of many worlds in this world'.[9]

By comparison with the regular appearance of snail-related material in premodern manuscripts, incunabula and elsewhere (as I'll discuss further on in this section), Shakespeare's works come late to the topic. In terms of premodern creative writing, Geoffrey Chaucer's epic verse *Troilus and Crisyede* (c.1380), for instance, contains depictions of its eponymous lovers as startled snails recoiling from the world.[10] Caroline Spurgeon's important 1935 study *Shakespeare's Imagery and*

What It Tells Us includes a page on 'his sympathy for snails', which illustrates how three quotations from *Venus and Adonis*, *Love's Labour's Lost* and *Coriolanus*—episodes that all focus Nashean-like on the snail's horns and movement—exemplify how they '[seem] to him an example of one of the most delicately sensitive organisms in natures'.[11] Spurgeon's close textual analysis shows how 'Shakespeare was able to immerse himself compassionately in the suffering of a snail'.[12] But it's a moot point, I think, to attribute culturally dispersed gastropod tropes to an individual's specific point of view. Furthermore, 'a major problem with these examples... is that they are all similes. These suffering snails figure in the text as a means of expressing human suffering'.[13] Nonetheless, this common figure of comparison is helpful for thinking about Shakespeare's texts as expressions of a late sixteenth-century European society entangled in foregoing humoral sensibilities and forthcoming Cartesian selfhoods.

To be sure, Shakespeare's representation of these beings' 'contractility' (to cite Marianne Moore's brilliant 12-line verse 'To a Snail' (1924)) and their complex combination of soft/hard matter connects to the kinds of ambivalent ontologies so clearly dramatised in *Hamlet*'s vacillations about being and not being. And I'm not alone in thinking closely about snails in Shakespeare. François-Xavier Gleyzon's remarkable book-length study *Shakespeare's Spiral: Tracing the Snail in King Lear and Renaissance Painting* (2010) is essential reading for anyone interested in the topic. Impelled by Lear's Fool's riddling lines to the King that 'a snail has a house./... to put his head in; not to give it away to his daughters, and leave his horns without a case' (1.5.23–6), Gleyzon's generally post-structuralist analysis examines the significance of the 'spiral shell of the snail' in 'Renaissance literature and...iconography of the same period'.[14] Taking an orthogonal critical stance, this section examines how Shakespearean snails raise questions related to surfaces, selfhoods and interspecies somatics. Specifically, I wonder if Spurgeon's comments on Shakespeare's approach to snails as 'sensitive organisms' is quite right. Should the phrase be 'sensitive organs'? Gleyzon writes that the snail's

> sticky and wet body exudes shining saliva over its path, rolling up and unrolling alternatively, but there is also its horns which it extend or contracts in its own way, just like its body in its shell, bringing to mind a whole sexual symbolism.[15]

After Gleyzon, I can't help but notice the poem's and the plays' slime-free treatment of the snail's passage between the inside and the outside of its shell are movements which mimic human anatomical parts such as tongues and penises, those 'subtle and shifting borderline[s] between the body and its meanings'.[16] And now these snail/human correspondences have occurred to me, there seems to be key differences between Shakespeare's use of snails in *Venus and Adonis* and the plays.

As a gloss for Venus' emotional texture when she comes across Adonis' boar-slain body, the poem's narrator invokes a snail who 'Shrinks backwards in his shelly cave with pain' when their 'tender horns' are 'hit' (1033–4). Shakespeare's snaily-simile continues to describe how the withdrawn creature 'there, all smothered up, in shade doth sit, / Long after fearing to creep forth again' (1035–6), thus providing further insight into the goddesses' transformation from the sexual aggressor that starts *Venus and Adonis* to the motherly form that ends it. While Venus is allied with the retreating snail and its shell (and here it's useful to remember that in 1612 Aegidius Albertinus compared snail shells to female genitalia),[17] *Love's Labour's Lost* turns to its 'tender horns' to make a broader, intellectual claim about human emotion. With a plot concerned with a King and his three friends who swear an oath to give up women for three years to focus on their studies only to find that they are enamoured with a Princess and her three women companions newly arrived at the court, this early comedy questions its titular emotion, which is so often at the heart of this particular genre. In 4.3 as part of a homosocially driven lengthy speech designed to argue for both scholarly study *and* the love of women, Biron explains to the rest of the oath-sworn quartet how 'Love's feeling is more soft and sensible/ Than are the tender horns of cockled snails' (4.3.311–12). It's pretty much standard practice to mention John Keats' admiration for Shakespeare's use of snail imagery in *Venus and Adonis*. The mode and content of Biron's statement to his intellectually aspirational male companions anticipates the Romantic writer's comments about sublime aesthetics in a letter dated 8 April 1818 to the painter Benjamin Robert Haydon. Here Keats remarks on 'The innumerable compositions and decompositions which take place between the intellect and its thousand materials before it arrives at that trembling delicate and snail-horn perception of beauty'.[18] Though separated by over two hundred years of cultural differences, Keats' letter and Shakespeare's courtly character are linked by discussions about feelings and snail-sensibilities that take place between men.

Molluscan-related masculinities also crop up in the Jacobean Roman tragedy *Coriolanus*. In that play, Menenius of Rome compares Afidius (the general of the Volscian army) to a snail 'Who, hearing of our Martius' banishment, /Thrusts forth his horns again into the world, /Which were inshelled when Martius stood for Rome, /And durst not once peep out' (4.6.45–8). A great deal has been written about the erotic texture of Coriolanus' and Aufidius' relationship.[19] At the very least, their relationship is one of heightened homosociality. If the snail is used as a simile in Shakespeare's poem to describe Venus' withdrawal from the world, it's deployed in *Coriolanus* to define the opposite, the Volscian leader's propulsion into action. In obvious terms, there's something undeniably phallic about Menenius' representation of Aufidius as a creature 'thrust[ing] forth' animate projections which are anything but 'tender'. In this Roman tragedy, snail horns help to define fecund masculinities, but in other plays they are signs of different kinds of premodern manhood.

Produced between *Love's Labour's Lost* and *Coriolanus*, *As You Like It* extends Shakespeare's use of snails to articulate human sexuality. The complex fusion of genders assembled around the art of wooing at the end of 3.3 continues in Act 4 as Rosalind-disguised-as-Ganymede-disguised-as-Rosalind and Orlando take up their sexually charged dialogue. Arriving 'within an hour of [his] promise to meet' (4.1.37–8) for a lesson in love, the disguised Rosalind chides Orlando: 'Nay, an you be so tardy, come no more in my sight. I had as lief be wooed of a snail' (4.1.45–6). When Orlando queries 'Of a snail'? (4.1.47), Rosalind-Ganymede-Rosalind explains their preference:

> Ay, of a snail; for though he comes slowly, he carries his house on his head—a better jointure, I think, than you make a woman. Besides, he brings his destiny with him [...], horns, which such as you are fain to be beholden to your wives

for. But he comes armed in his fortune, and prevents the
slander of his wife. (4.1.48–54)

In anticipation of Lear's Fool's quip, which also aligns material possession, marital fidelity and molluscular figures, *As You Like It* calls to mind contemporaneous associations between snail horns and cuckoldry. Earlier in the same play, the melancholic philosopher Jaques turns to the prone aspect of the gastropod's body rather than its erect tentacles to describe prepubescent masculinity, 'the whining schoolboy [...] creeping like snail/ Unwillingly to school' (2.7.144–6). With Rosalind-Ganymede-Rosalind's and Jaques' quotations in mind, it's striking that *As You Like It* (a Shakespearean comedy known for blurring fixed lines of gender demarcation) ties the creature known since John Ray's observation in 1660 for hermaphroditic reproduction to masculine sexualities.[20]

Before Ray's mid-seventeenth-century malacological discovery, snails were generally discussed in terms of their relationship to material surfaces and spiritual sensibilities. The description of Jaques' schoolboy 'creeping like snail' seems alert to the broadly pejorative Christian view which constructs a hierarchical binary opposition between upright, bipedal humans and beings who enjoy a more horizontal relationship with the ground, that is 'every creeping thing of the earth'.[21] And at either end of the human's lifecycle, like Jaques' schoolboy or the aged lover in George Chapman's comedy *May Day* (1611) who dresses up as a character called Snail to pursue the young woman who is the object of his sexual desire, youth and senescence are figured in apparently deficient ways. Combined with its horizontal demeanour, premodern Christianity's notion of the snail's spontaneous generation such as Bartholomew the Englishman's fourteenth-century Franciscan notion that it reproduces in 'lime or of lime and is always foul and unclean' added to the organism's subordinate position.[22]

On the other hand, Huguenot artisans fleeing France for England and America during that country's civil wars of religion in the mid-sixteenth century used snail imagery to bring forth a specific kind of Protestant spirituality. Underpinned by the Calvinist writings and praxis of potter, ceramicist and proto-scientist Bernard Palissy, Pallissian piety viewed the snail as an emblematic representation of 'the fortress city' described in the artisan's *La Recepte Véritable* (1563) and summarised here in Maria Jolas' translation of Gaston Bachelard's *The Poetics of Space* (1958):

> Faced with 'the horrible dangers of war,' Bernard Palissy contemplated a design for a 'fortress city.' He had lost all hope of finding an existing plan 'in the cities built today.' Vitruvius himself could be of no help in the century of the canon. So he journeyed through 'forests, mountains and valleys to see if he could find some industrious houses.' After inquiring everywhere, Palissy began to muse about a young slug that was building its house and fortress with its own saliva. Indeed, he passed several months dreaming of a construction *from within*... Meanwhile, someone brought Palissy two large shells from Guinea: 'A murex and a whelk.' The murex being the weaker must be the best defended, according to Palissy's philosophy.[23]

Palissy's scrutiny of the tropical sea snail paved the way for an imagined city-scape which starts to take shape from an open square housing the city's governor. All the other houses (their windows and doors facing inwards to the governor's) would be built in a single street winding around that central dwelling. The backs of the houses made one single wall. 'The last of the house-walls was to back up against the city wall which, thus, would form a gigantic snail'.[24] But as Bachelard/Jolas makes clear, there's more to Palissy's snail-inspired city than architectural nous.

Instead of handling snail generation with suspicion, Palissy's crafts of pottery and ceramics informed his perception of 'these secreted molluscan fortresses as portable wonders'.[25] Like the potter, snails made something solid from liquid materials. But more than this, and as Neil Kamill has shown, 'Palissy observed that the common mollusc constructed a portable fortress from hidden interior resources and carried it everywhere on its back, as did the pious Huguenot artisan his craft knowledge and tools'.[26] It's quite remarkable to think that at about the same time Palissy was meditating on the material and religious significance of snails that England was developing another gastropod genre.

Shakespearean texts use snails to define human activities foreground in the poem's and plays' plots. Their bodies are fragmented into individual parts such as 'shelly caves' or 'horns'. They have a singular mode of movement: slow. Audiences encounter a very different snail in the mid-sixteenth-century court interlude *Thersites* (published in 1562 and attributed to Nicholas Udall). Although the snail doesn't feature on the title-page's list of 'The names of the players', a snail takes the stage at line 388. The stage directions state 'Here a snaile muste appere unto [Thersites] and hee muste loke fearefully uppon the snaile'.[27] According to Marie Axton, '*Thersites* has plenty of action but little plot'.[28] The play begins with the cowardly braggard of the play's title getting ready for war and much of the early action involves Thersites fighting the snail.

Thersites dramatizes those intriguing battles between snails and knights that populate medieval manuscripts' marginalia and which continue to lurk in twenty-first century popular culture, for example in the British Library's blog entry 'knight v snail'[29] and the Radio 4 programme *Knight Fights Giant Snail*.[30] The iconography of skirmishing knights and snails goes beyond illuminated manuscripts. A historiated column inside the Romanesque priory at Romainmôtier-Envy, Nord in Switzerland features one such clash. The European context is relevant for the vernacular interlude. Mark Albert Johnston tells us that:

> By the middle of the sixteenth century, humanist schoolmasters began translating classical Roman interludes into English for performance by their students: ...*Thersites* was translated from an early sixteenth-century French text composed in Latin—likely by Nicholas Udall, either in his early years at Oxford or while headmaster at Eton—and performed either at court or in Oxford during Henry's reign.... In medieval literature and manuscript illumination, the snail traditionally symbolises cowardice; so when Thersites treats the slug as a substantial threat, the parody demonstrates his considerable distance from the ideal of martial manhood: ... the snail typifies behaviour that anticipates precisely Thersites's own response to the combative challenge posed later in the play by the bearded soldier Miles.[31]

Although it's not entirely clear what all these medieval snails and knights represented,[32] in Johnston's analysis the mollusc stands in for human cowardice and impoverished 'martial manhood'. Another comparative study states that:

> Published in 1562 as *A new Enterlude called Thersytes*, the comedy is an elaborate expansion of an earlier Latin dialogue by the influential French humanist, Ravisius Textor. Also entitled *Thersites* (and printed posthumously in Paris, 1530), Textor's 267-line dialogue is increased to over 900 lines plus stage directions. The comprehensive additions and elaborations in the English version include Thersites' opening soliloquy: while Textor's dialogue begins with Thersites demanding armour from Vulcan (Mulciber in Udall), the vernacular Thersites addresses his audience, introducing not only himself but also a pervasive metatheatricality, and explicit intertextuality, that is scarce in the neo-Latin source.[33]

There's something distinctive, then, about this vernacular rendition of the snail. It's worth noting that Textor's Latin Thersites' non-human opponent (and even though it has horns like a snail) is referred to as testudo ['tortoise'].[34] Given that 'testudo' is the term used to describe how Roman soldiers overlap their shields to form an overhead cover from enemy weapons, Textor's word-choice might enable a slightly different response to Thersites' cowardice than its English counterpart. Nonetheless, with Johnston's and Kenward's analyses in view, the domesticated interlude clearly uses 'Snail' to say something about its titular classical protagonist. Thersites' interest in staging snails thus foregrounds humanism's narcissistic engagement with the world, that is its desire to see humankind's traits reflected back to itself.

In so doing, and as exemplified in the interlude's closing lines, non-human matters help establish a hierarchical Christian social order:

> Love God, and feare him, and after him youre kinge
> Whiche is as victorious as anye is lyving.
> Praye for His Grace, with hartes that dothe not fayne,
> That longe he maye rule us withoute grefe or paine.
> Beseche ye also that God maye save his quene,
> Lovelie Ladie Jane, and the prince that he hath send
>   them betwen
> To augment their joy and the comons felicitie.
> Fare ye wel, swete audience, God graunt you al prosperite.
> Amen. (908–15)

In line with humanist critical conventions, scholarly accounts of *Thersites* discuss its central main character's classical analogue in Homer's *Iliad*, continental source text and courtly audience. But what happens if our discussion resists its obvious humanist agenda? What if our critical focus begins with Snail, a cast member so clearly missing from the title page's list of characters? What other views of the play might be available if we consider the other non-human aspects of this curious Tudor entertainment? I first want to suggest that looking beyond that braggart soldier and its scholarly background allows for a discussion of Thersites' interests in broader creaturely environments befitting a humoral as much as a humanist agenda. Next, I want to consider a kind of snail sensibility, which tries to think *with* rather than beyond non-human figures.

Thersites' humoral agenda begins in the comic (if overplayed) phonic slippage between sallet/helmet and sallet/herbs in the opening dialogue between Thersites and Mulciber a Smith. Mulciber's side comment that he 'eate none suche sallettes for now I waxe olde / And for my stomacke they are verye coulde' (42–3) is suggestive of the smith's phlegmatic disposition and one which (according to humoral theories) should avoid cold food. Mulciber's humoral attitude is thus quite different from Thersites' more choleric activities. Yet both humans are corporeally weak. The smith's physical vulnerability is shown through dietary requirements: Thersites' physical vulnerability is shown by his need for Mulciber's protective armour in the first place.

The interlude's pre-Cartesian world shows non-human and human agents enmeshed and Thersites' encounter with the snail shows that enmeshment quite literally. The episode begins with the telling stage direction mentioned above. Thersites then says:

> But what a monster do I see nowe
> Comminge hetherwarde with an armed browe?
> What is it? Ah, it is a sowe!
> No, by Gods body, it is but a grestle
> And on the backe it hath never a brystle.
> It is not a cow—ah there I fayle
> For then it should have a long tayle.
> What the devyll! I was blynde—it is but a snayle.
> I was never so afrayde in east nor in south;
> My harte at the first syght was at my mouth.
> Mary syr, fy, fy, fy! I do sweate for feare—
> I thoughte I had craked but to tymely here.
> Hens thou beest and plucke in thy hornes
> Or I sweare, by him that crowned was with thornes
> I will make [thee] drincke worse than good ale in the cornes.
> Haste thou nothynge elles to doo
> But come wyth hornes and face me so?

> Howe, how, my servauntes, get you sheld and spere
> And let us werye [make war] and kyll thys monster here.
>
> (388–406)

Of course, Thersites' pusillanimity and garrulity matches the diminutive aspect of his silent opponent, although there's nothing to suppose that the snail is true-to-size. (Those of us familiar with *Monty Python and the Holy Grail* (1975) might be reminded of its 'Killer Rabbit of Caerbannog' as we read this Tudor interlude.) Whatever the snail's dimensions, Thersites calls attention to its physicality in ways that Shakespeare obviously avoids. At the same time, the interlude's humour depends on the cultural consensus on what premodern snails mean. We have already seen the rich valency of snail horns in Shakespearean texts and their connection to masculine sexuality. But alongside those understandings, Thersites' focus on the approaching non-human body, which is 'but a gristle', is also briefly interested in the texture of its corporeal surface too. Thersites' call to his servants to take up arms shows that he doesn't want to fight the creature alone. The comedy is thus increased as this impassive snail is juxtaposed with an increasingly agitated Thersites who is joined onstage by Miles ('a pore souldiour come of late from Calice' (411)) just after he calls for aid.

The scene is punctuated by Thersites' taunting of the snail, Miles' goading of Thersites and a mute mollusc which reacts with minimal physicality. Three successive stage directions tell us: 'Then [Thersites] must fyghte against the snayle with his club'; 'And [Thersites] must cast his club awaye'; 'Here [Thersites] must fighte then with his sworde against the snayle, and the snayle draweth her hornes in' (444–55). I've often wondered if there's any significance in the stage directions gendering the snail as female. I'm assuming it's primarily because 'testudo' (the word used in the source text remember) is a feminine noun. But given the interlude's treatment of Thersites' martial masculinity—and recalling Shakespeare's treatment of snails—I can't help but think that the sexual politics of the staged snail/knight battle is relevant too: it adds to the comedy if this is a specifically *female* mollusc.

Once the snail recedes from narrative view (though the creature may well remain onstage) the interlude becomes focused on another kind of physical encounter between non-human and human organisms: the child Telemachus' intestinal infection caused by parasitic worms. Instead of weaponry, Thersites' mother is enlisted—under threat by her son—to cure the boy's malady through a lengthy charm that is simultaneously entertaining and disturbing. This domestic scene offers a parallel to Thersites' combat with the snail that opens the play. While the worm's threat to the child comes from within not without the human body, both Thersites' snail and worm threaten masculine bodies.

Before the age of microscopy and John Ray's patient study which revealed the mollusc's non-binary status, the Tudor interlude stages the snail in a remarkable way. *Thersites* asks its audience not to think of man-as-snail in the way that the early seventeenth-century painting 'Snail Man' (attributed to Jacques Callot), for example, demands. Rather, it nudges its spectators to look past the human and to think of snail-as-snail. I've suggested throughout *Shakespeare on the Ecological Surface* that premodern humoral thinking connects interiority and exteriority and troubles a layering of clearly demarcated surfaces. In *Thersites*, human characters share the stage with a snail. And once you know about the gastropod's agency in this mid-sixteenth-century play then Shakespeare's snails look increasingly anthropomorphic in their ecological outlook.

Like many others, some of whom I've mentioned in this section, I'm intrigued by snails and I'm particularly interested in the creative and critical thinking these living things inspire.[35] Frances Ponge's 1942 prose poem 'Snails' speaks to their exemplarity. They are, writes Ponge,

> saints who make masterpieces of their lives, works of art of their own perfection....How are they saints? Precisely by their obedience to their nature. So: know yourself. And accept yourself for what you are. In agreement with your vices. In proportion with your measure.[36]

Snails are just one non-human being that humans use to construct their authoritative place in a hierarchical social environment. If humans push against the grain of Shakespearean simile towards a different mode of correspondence—humans are not like snails but rather humans live *alongside* these molluscular creatures and can learn from them—then there might be a way that all species can exist 'in proportion with [their] measure'.

NOTES

1 Elisabeth Tova Bailey, *The Sound of a Wild Snail Eating* (Devon: Green Books, 2010), p. 13.
2 Bailey, *The Sound of a Wild Snail Eating*, p. 4.
3 Bailey, *The Sound of a Wild Snail Eating*, p.12, p.32.
4 Dusita Saokaew, *China Global Television Network* (CGTN), 'milking "happy" snails for beauty products', 21 September 2019. https://news.cgtn.com/news/2019-09-21/Milking-happy-snails-for-beauty-products-K9TSfS7P4Q/index.html.

5 Gaby Del Valle, 'The ethics of snail mucin', *The Outline*, 11 May 2018. https://theoutline.com/post/4503/snail-mucin-farms-extraction-skin-care-heliciculture-ethics.
6 Thomas Nashe, *The Unfortunate Traveller* (1594), C2r. I would like to thank Helen Davies for alerting me to this quotation.
7 Ronald Chase, 'Lessons from snail tentacles', *Chemical Senses*, 11.4 (1986): 411–26, p. 413.
8 Patricia Highsmith, 'The snail-watcher', *Eleven* (New York: Atlantic Monthly Press, 1970), pp. 3–10, p. 10.
9 I'm citing Margaret Cavendish's poem of that name from *Poems and Fancies* (1653), p. 44.
10 Elizabeth Robertson, 'First encounter: "Snail-horn perception" in Chaucer's *Troilus and Criseyde*', in *Contemporary Chaucer across the Centuries*, edited by Helen M. Hickey, Anne McKendry and Melissa Raine (Manchester: Manchester University Press, 2019), pp. 24–41, p. 24. I am very grateful to Elizabeth Robertson for sharing a pre-publication version of her essay with me.
11 Caroline Spurgeon, *Shakespeare's Imagery and What It Tells Us* (Cambridge: Cambridge University Press, 1935), p. 107.
12 Kristine Steenbergh, '"The tender horns of cockled snails": Feeling with vermin in early modern England', Society for Renaissance Studies [SRS] Conference Sheffield 2018—Panel 9: Emotions and the Natural World in Early Modern Culture, Jessop West SR8. Chair: Shani Bans (UCL). I am very grateful to Kristine Steenbergh for sending me a copy of her SRS talk and providing permission to cite from the unpublished talk.
13 Steenbergh, '"The tender horns of cockled snails"'.
14 François-Xavier Gleyzon, *Shakespeare's Spiral: Tracing the Snail in King Lear and Renaissance Painting* (Maryland: University Press of America, 2010), p. xii.
15 Gleyzon, *Shakespeare's Spiral*, p. 3.
16 The quotation comes from Louise O. Fradenburg's discussion of the tongue and the twelfth-century *Sefer Zekirah, The Book of Remembrance of Rabbi Ephraim of Bonn* in the article 'Criticism, anti-semitism and "The Prioress' tale"', *Exemplaria*, 1.1 (1989): 69–115, p. 80.
17 Anna Grasskamp, 'Shells, bodies, and the collector's cabinet', in *Chonchophilia: Shells, Art, and Curiosity in Early Modern Europe*, edited by

Marissa Anne Bass, Anne Goldar, Hanneke Grottenboer and Claudia Swan (Princeton, NJ: Princeton University Press, 2021), pp. 49–74, p. 65.

18 Sidney Colvin (ed), *The Letters of John Keats* (London: Macmillan, 1925), A Project Gutenberg eBook. www.gutenberg.org/files/35698/35698-h/35698-h.htm#XLVII p. 95. I was alerted to this letter in Roberston, 'First encounter: "Snail-horn perception" in Chaucer's *Troilus and Criseyde*', 2019, pp. 33–4.

19 See, for example, Stanley Wells, *Shakespeare's Tragedies: A Very Short Introduction* (Oxford: Oxford University Press, 2017), p. 114–15.

20 See John Ray, *Catalogus plantarum circa Cantabrigiam nascentium* (1660) as discussed in Aydin Örstan, 'John Ray's hermaphrodite snails on their 350th anniversary', *Mollusc World*, 23 (2010), p. 3.

21 Geneva Bible, Genesis 1:25.

22 Karl Steel, *How Not to Make a Human: Pets, Feral Children, Sky Burial, Oysters* (Minneapolis: University of Minnesota Press, 2019), p. 97.

23 Gaston Bachelard, *The Poetics of Space*, translated by Maria Jolas (Boston, MA: Beacon Press, 1994), pp. 127–8.

24 Bachelard, *The Poetics of Space*, p. 129.

25 Neil Kamill, *Fortress of the Soul: Violence, Metaphysics, and Material Life in the Huguenots' New World, 1517–1751* (Baltimore, MD: Johns Hopkins University Press, 2005), p. 76.

26 Kamill, *Fortress of the Soul*, p. 5.

27 Marie Axton (ed), *Three Tudor Classical Interludes* (Cambridge: D. S. Brewer, 1982), pp. 5–15, pp. 37–63. Quotations from *Thersites* are from this edition. Line references are provided parenthetically.

28 Axton, *Three Tudor Classical Interludes*, p. 5.

29 'Medieval manuscript blog', The British Library, 26 September 2013. https://blogs.bl.uk/digitisedmanuscripts/2013/09/index.html.

30 Alixe Bovey, *Knight Fights Giant Snail*, BBC Radio 4, 28 May 2020. www.bbc.co.uk/programmes/m000jgfv.

31 Mark Albert Johnston, *Beard Fetish in Early Modern England* (London: Routledge, 2016), pp. 122–5.

32 Michael Camille, *Image on the Edge* (London: Reaktion Books: 1992), pp. 31–6.

33 Claire Kenwood, '"Of arms and the man": Thersites in early modern English drama', in *Epic Performances from the Middle Ages into the Twenty-First Century*, edited by Fiona Macintosh, Justine McConnell, Stephen

Harrison and Claire Kenward (Oxford: Oxford University Press, 2018), pp. 421–38, pp. 421–2.
34 Textor's Latin describes the creature as 'geminis testudo' [twin tortoise], although Marie Axton notes that 'Textor's hero never names his beast'. Axton (ed), *Three Tudor Classical Interludes*, p. 150, p. 170n.T395.
35 See, for example, my X hashtag #AdoreSnails.
36 Francis Ponge, 'Snails', in *Partisan of Things*, translated by Joshua Corey and Jen-Luc Garneau (Chicago, IL: Kenning Editions: 2016), pp. 18–21, p. 21.

Silk

Textile production

Eight

The popularity of silk face masks as a defence from the Covid-19 virus at the start of the global pandemic brought this protein-fibre fabric to the attention of the general public in new ways. In the spring of 2020, the media's coverage of 'fashion in the time of Covid'[1] usually included discussions of face masks made from the kind of commercial silk derived from Bombyx moths' caterpillar cocoons. While some saw these masks as more of a sartorial choice than a safeguarding decision, later research told us how 'Silk fiber has antimicrobial properties. And silk is hydrophobic, which means it sheds water unlike cotton masks that typically absorb it', thus a double layer of silk improves the function and extends the lifespan of surgical masks.[2] Moreover, silk masks' ability to reduce the 'chemical pollutants and nano-plastics'[3] unleashed by its single-use disposable counterpart is as welcome as their capacity to repel infection and moisture. Long before Covid-19's potentially deadly droplets shed fresh light on its remarkable properties, silk's global history is linked to various promising and perilous human endeavours.

In common with so many of the nouns considered in this book, silk is a wonderfully mutable substance. 'Silk', we're told, 'begins as a liquid extruded by a caterpillar through a small opening on its back. Upon contact with the air, the liquid solidifies into a filament. This filament constitutes

DOI: 10.4324/9780429326752-9

the raw material for a highly coveted textile'.[4] The trope of transformation that characterises the shift in the caterpillar's excretion from fluid to fibre is extended by the further metamorphosis of that fibre into a luxurious fabric:

> Ever since Antiquity, the love and longing for silk textiles has fostered connections across Asia, Europe, and the Atlantic. By the late sixteenth century the trade of silk yarn and cloth was truly global, spanning all known continents. Many strove to discover the principles of silkworm breeding and unveil the secrets of producing high-quality silk yarn. Sericulture and silk manufacturing thus propelled both peaceful interchange and cross-cultural competition, generating wealth and power, and transforming economies and societies.[5]

Like the literal warp and weft which eventually structures a silken surface, silk manufacture is characterised by 'peaceful interchange and cross-cultural competition', discourses of human interaction which weave their way through early English texts. What's often marginalised in even the most capacious accounts of silk's history is the invaluable part taken by the tiny non-human beings that make all this possible.

As part of a critically agile and wide-ranging analysis of how Shakespearean texts '[entwine] trees, silkworms, silk (derived from mulberry leaves), and the human, anticipating Bruno Latour's theory of the assemblage', Todd`A. Borlik shows how Shakespeare's 'writing…conveys a keen appreciation of the powers early modern culture invested in this fabric'.[6] For Borlik, 'an eco-materialist study of the mulberry and its afterlife' foregrounds 'Shakespeare's evident fascination with silk' and 'reflects his acute understanding, as both a glover's son and an actor, of the ways in which human subjectivity is enmeshed within nonhuman materials'.[7] Citing the line from *King Lear* 'thou owest the worm no silk' (Quarto 11.88) as just one example, Borlik points out that if many 'neglect to consider the labor of nonhuman agents in the cloth economy' then 'Shakespeare reminds us of it'.[8] In several ways, my ensuing comments on textilic power and prestige in Shakespearean works are indebted to Borlik's study of the period's and the playwright's 'fetishizing of silk as a marker of aristocratic identity'.[9] However, we differ in our summative outlooks. Borlik's understanding of 'Shakespeare and the sylk man' concludes 'it is not the texture but the *sound* of the silk that Shakespeare finds erotic'.[10] To be sure, Shakespeare doesn't match Robert Herrick's depiction of the sensuous fabric in his 1648 lyric 'On Julia's Clothes'. That poem's narrator ponders:

When as in silks my *Julia* goes,
Then, then (me thinks) how sweetly flowes
That liquefaction of her clothes.
Next, when I cast mine eyes, and see
That brave Vibration each way free;
O how that glittering taketh me![11]

Here, Herrick's use of the word 'liquefaction' captures the alchemical quality of silk's lustrous surface rendered fluid by Julia's animated body beneath. I'm interested in how Shakespeare's apparent reticence to consider silk's surficial glamour eventually takes shape. Following strands of sericultural matter in *A Midsummer Night's Dream*, *The Comedy of Errors*, *The Merchant of Venice*, *Othello*, *The Winter's Tale* and Kenneth Branagh's 2006 film adaptation of *As You Like It*, this section asks if Shakespeare can help make the silkworm's role in silk's textual production even more visible than Borlik suggests.

By no means a word defining a homogeneous fabric,[12] silk's importance for late sixteenth- and early seventeenth-century England is most clearly shown not by Shakespeare but by Thomas Moffett in his 76-page two-book poem first published in 1599 (but 'penned perhaps as early as 1589').[13] Dedicated to the well-known literary patron Mary Sidney, described here as 'the most renowned Patronesse, and noble Nurse of Learning', the full title of Moffett's work *The Silkewormes, and their Flies: Lively described in verse, by T. M. a Countrie Farmar, and an apprentice in Physicke. For the great benefit and enriching of England* speaks to the poet's and the nation's growing sense of its silk economy.[14] Taking up a Homeric persona who will 'sing of little Wormes and tender Flies,/ Creeping along, or basking on the ground' in the text's dedicatory poem,[15] the poet casts his addressee as the 'Sydneian Muse' who will 'help [Moffett] sing these flocks as white as milke,/ That make, and spinne, and die, and windle silke'.[16] Under the guise of a typical Arcadian riff, and framed by a titular woodcut showing the caterpillar, the fly and the cocoon and prefaced by a Table comprised of 30 topics (from 'When garments were first used' to 'Keeping of Silke-wormes hindereth neither Shepheardes, Spinsters, Weavers, nor Clothiers'),[17] Moffett provides a pastoral depiction of the creatures' lifecycle from birth to the making of the threads that form a silken surface. Indeed, Moffett's learned yet engaging verse (as the title observes it's 'lively') is clearly connected to the georgic world of labour and material consumption:[18]

> For sure I know thy knowledge doth perceive,
> What breth embreath'd these almost thingles things:

> What Artist taught their feete to spinne and weave:
> What workman made their slime a robe for kings,
> How flies breed wormes, how wormes do flies conceive:
> From natures womb, how such a nature springs,
>> Whereof none can directly tell or reade,
>> Whether were first, the flie, the worme, or seede.[19]

The Silkewormes, and their Flies makes its material case for silk manufacture quite lightly. One of the most direct comments on silkworm cultivation appears towards the end of the poem as a marginal note, which guides the reader to 'An exhortation to all Farmers and Husbandmen to plant Mulberries'.[20] A few pages later, the poem's generally genial tone is momentarily replaced by a rousing couplet crying 'Up Britaine blouds, rise hearts of English race, / Why should your clothes be courser than the rest?'.[21] Most of the time, Moffett's poem is a fine example of how a Christian-humanist outlook combines sacred and secular worldviews while simultaneously navigating asymmetrical power relations between the poem's patron and its patronised writer, and human and non-human beings.

After the first few verses, the poet's initial deference to his dedicatee's knowledge of these 'almost thingles things' gives way to a rewriting of the Fall, which brings in the spiritual as well as the mundane need for clothes and how those material needs are serviced by *The Silkewormes, and their Flies*. In so doing, Moffett 'makes the crucial insight that his discussion of silkworms is itself affected by the legacy of original sin'[22] and thus Moffett's poem declaims,

> The breast which yet had hatcht no badde conceat,
> Nor harbour'd ought in heart that God displeaz'd,

> Did it for silken wastcoates then intreate?
> Sought it with Tyrian silks to be appeaz'd?
> No, no, there was no neede of such a feate.[23]

Moffett's verse recalls how postlapsarian humankind wore 'skinnes of beasts, (to shew their beastly fall)' and silk (along with flax) arrives after 'the floud had sinners swept away'.[24] In another time and place, Jacques Derrida reminds us that 'In principle, with the exception of man, no animal has ever thought to dress itself. Clothing would be proper to man, one of the "properties" of man'.[25] As a by-product of a worm rather than the surface covering of a 'beast', silk becomes allied with, not separated from, being human. And over time, as we'll see below, silk's creaturely origin almost disappears.

Based on 'Love's Schoolmaster' work,[26] that is Ovid and his story of 'Pyramus and Thisbe' from Book 4 of the *Metamorphoses*, one of Shakespeare's (and early modern England's) most popular poems in translation,[27] *The Silkewormes, and their Flies* constructs a 26-stanza aetiology of silk. Whereas Ovid's tragic story ends with the dead lovers' blood turning the mulberry tree's fruit from white to purple, Moffett extends the Ovidian material to suit his text's specific focus on the silkworms:

> Since which time eke some other Authors faine,
> Their humming soules about these haplesse trees,
> To be transported from th'Elysian plaine,
> Into the snowy milke-white Butterflyes:
> Whose seedes when life and mooving they obtain,
> How e're they spare the fruit of Mulberies,
>   Leave yet no leaves untorne that may be seene,
>   Because they only still continue greene.

...
When leaves are gone, and summer waning is,
The little creepers never cease to move,
But day and night (placing in toyle their blisse)
Spinne silke this tree beneath and eke above:
 Leaving their ovall bottoms there behind,
 To shewe the state of ev'ry Lover's mind.[28]

With a marginal reference to the last Book of Natale Conti's *Mythologiae* (1567), which speaks of how 'these stories [inform] us that the world was created by God, and that it was made from eternal matter',[29] Moffett refers to 'other Authors'' stories of the transformation of Pyramus and Thisbe's 'humming souls' into 'snowy milke-white Butterflyes'. This material functions as a bridge between Ovid and Moffett's own intense scrutiny of the silkworm's development, from the gluttonous 'little creepers' who 'spinne silke' to the airborne beings who leave 'their ovall bottoms [cocoons]' about the mulberry tree. It's at this point that Shakespearean surfaces intersect with those in Moffett's poem.

Sericultural and Shakespearean scholars often link Moffett's word for cocoon (the first time 'bottom' is used in this context according to the OED)[30] to *A Midsummer's Night's Dream's* weaver Nick Bottom and to Ovid's myth of Pyramus and Thisbe,[31] the play-within-a-play that Bottom takes part in so enthusiastically and erroneously: he famously refers to Ovid's blood-soaked story as 'this comedy of Pyramus and Thisbe' (3.1.8–9). In these respects, Bottom is probably the most silk-suffused character in the Shakespearean canon. Puck's reference to the entire artisanal acting troupe as 'hempen homespuns' (3.1.65), however, draws attention to the cultural, social and economic disparity between the silk fibres suggested by Nick Bottom's name and the fabric of the clothes he wears. If Moffett's poem addressed to the aristocratic Mary Sidney poetically probes silk's enigmatic qualities, then Puck's collective term for *A Midsummer Night's Dream's* labouring folk speaks to hemp's ubiquity in a nation urged by statutes to propagate this agricultural crop.[32]

England's hierarchical approach to textiles is writ large in the Tudor sumptuary laws, which set boundaries for who could wear what fabric 'in respect of their colour, quality, quantity, price, and make, on a graduated basis according to the condition and means of the wearer'.[33] One of the most well-known Shakespearean responses to such Elizabethan laws is the on-stage fall-out caused when *Twelfth Night's* socially aspirational steward Malvolio is duped into wearing cross-gartered yellow stockings. While it's not the stocking's fabric that leads to Malvolio's humiliation, and as Maria knows all too well, these textile objects have influence in Countess Olivia's household. The yellow stockings play a key part in what Malvolio calls 'the most notorious geck and gull/ That e're invention played on' (5.1.332–33)—and what twentieth-first century Anglo-American cultures might call gaslighting[34]—which results in his exit from both his job and the play. In a less dramatic but equally class-inflected scene involving silk stockings, while *Henry IV's Part Two's* Prince Hal's regal status should make him unaware of 'how many silk stockings' the 'irregular humorist' Ned Poins owns, the fact that the Prince does notice that Poins is currently wearing 'peach-coloured ones' (2.2.14–15) foregrounds the garment's role in the construction of sixteenth-century social hierarchies.

Henry VIII's 1510 'An Act against wearing of costly Apparel', for example, states that:

> no person under a knight (excepting sons of lords, judges, those of the king's council, and the mayor of London) is to wear velvet in his gown and doublet, or satin or damask in his gown or coat; and no person (with certain exceptions) not possessing freeholds to the yearly value of £20 may wear satin or damask in his doublet, or silk or camlet in his gown and coat.[35]

Later Elizabethan Statutes of Apparel and the 'eight proclamations' on the theme of 'excess of apparel'[36] were upheld by draconian measures including a November 1559 proclamation proposing

> that two watchers should be appointed for every [London] parish, armed with a schedule of all persons assessed to the late subsidy at £20 per annum, or £200 in goods and upwards, in order to see that the prohibition against silk trimmings was being obeyed.[37]

In the 1560s, men's hose was singled out for particular surveillance. According to a 1562 proclamation, 'Neither any man under the degree of a Baron, to wear within his hosen any velvet, Satin or any other stuff above the estimation of Sarcenet, or Taffata'.[38] By contrast with its 'homespun' counterpart, silk's legal standing alongside England's inhospitable climate for the textile's propagation increased its cultural and economic value.

A decade or so after Nick Bottom's suggestive sericultural appearance in *A Midsummer Night's Dream*, *The Winter's*

Tale stages an economically charged episode via Autolycus' Jacobean version of an advertising jingle:

> Will you buy any tape,
> Or lace for your cape,
> My dainty duck, my dear-a?
> Any silk, any thread,
> Any toys for your head,
> Of the new'st and fin'st, fin'st wear-a?
> Come to the pedlar,
> Money's a meddler,
> That doth utter all men's ware-a. (4.4.302–10)

Autolycus trades in cheap off-cuts of 'the new'st and fin'st' fabrics, including but not exclusively silk, to tempt *The Winter's Tale*'s pastoral population. The character's mercantilist approach to apparel is amplified in the song's nifty homophonic pairing of 'wear' and 'ware'. When compared to Autolycus' low-level monetary operation, silk has a key role in *The Comedy of Errors*' and *The Merchant of Venice*'s stage-worlds of global trade. In an opening scene which jars with the farce-like action that follows, the dialogue between Egeon and Solinus, Duke of Epeshus exposes the material anxieties bound up with Elizabethan trade:

> *Enter [Solinus], the DUKE of Ephesus, with [EGEON] the*
>   *Merchant of Syracuse, JAILER, and other attendants*
> EGEON Proceed, Solinus, to procure my fall,
> And by the doom of death end woes and all.
> DUKE Merchant of Syracusa, plead no more.
> I am not partial to infringe our laws.
> The enmity and discord which of late

> Sprung from the rancorous outrage of your Duke
> To merchants, our well-dealing countrymen,
> Who, wanting guilders to redeem their lives,
> Have sealed his rigorous statutes with their bloods,
> Excludes all pity from our threat'ning looks.
> For since the mortal and intestine jars
> 'Twixt thy seditious countrymen and us,
> It hath in solemn synods been decreed,
> Both by the Syracusians and ourselves,
> To admit no traffic to our adverse towns. (1.1.1–15)

The Duke explains that if any Ephesian is seen in Syracuse and vice-versa, they are subject to a fine of 'a thousand marks' (1.1.21) in goods or money. Egeon cannot raise the required sum 'Therefore by law [he is] condemned to die' (1.1.25). Anyone wandering into the start of Gray's Inn performance on 28 December 1594[39] without prior knowledge of *The Comedy of Error*'s Plautine-inspired plot might have been forgiven for thinking that a tragedy was about to unfold.

The Comedy of Error's opening scene calls attention to the kinds of human suffering caused by the 'laws' and 'statutes' of international trade. It turns out, however, that the Duke's curiosity about why the merchant 'cam'st to Ephesus' (1.1.30) saves Egeon and the play's generic status. The merchant of Syracuse's detailed account of his marriage, the birth of his identical twin sons, the 'mean-born' (1.1.54) identical twin sons he 'bought, and brought up to attend [his sons]' (1.1.57), and the shipwreck dividing them earns a day's reprise from the Duke. Paradoxically, Egeon's ability to 'tell sad stories of [his] own mishaps' (1.1.120) sets up the transition from the play's tense start to the on-stage mayhem caused when Antipholus of Syracuse and his bondsman Dromio of Syracuse end up in Ephesus, the home of their unknown respective twin brothers Antipholus and his bondsman Dromio. Given Act 1 Scene 1's co-mingling of servitude, privilege and power, it's hardly surprising that the considerable tensions on show at the start of *The Comedy of Errors* never go away. Though brief, silk's irruption three acts later works as a spur for highlighting how identities are ideologically fashioned and, quite literally, wrapped up in fabric.

In Act 4 Scene 3 Antipholus of Syracuse's confusion is depicted via a sumptuous soliloquent scene recalling Tudor legislation:

> There's not a man I meet but doth salute me
> As if I were their well-acquainted friend,
> And every one doth call me by my name.
> Some tender money to me, some invite me,
> Some other give me thanks for kindnesses.
> Some offer me commodities to buy.
> Even now a tailor called me in his shop,
> And showed me silks that he had bought for me,
> And therewithal took measure of my body.
> Sure, these are but imaginary wiles,
> And Lapland sorcerers inhabit here. (4.3.1–11)

Of course, the audience know that there's nothing supernatural at work here. Despite taking 'measure of [his] body', the Ephesians are just mistaking one identical twin for the other, treating their supposedly wealthy merchant neighbour fittingly and acknowledging Antipholus of Ephesus' right to wear the valued textile. Whereas *The Comedy of Errors'* invokes a luxuriously-adorned masculine body, *The Merchant of Venice* conjures a fabric-dressed surface of another kind. Set in a geographical location which allowed 'Venetian aristocracy' to make 'its fortunes.... from trade in merchandise imported to their city from every region of the known world, especially the kingdoms of Asia: China, the East Indies, Egypt, Syria, and Turkey',[40] silk helps convey *The Merchant of Venice*'s fraught economic environment at the outset. Responding to Antonio's impalpable sadness which launches the play at 1.1.1, Salerio gives an account of a shipwrecked mercantile venture which would 'Enrobe the roaring waters with [his] silks' (1.1.34): an initially captivating yet ultimately ruinous image which—like the play at large—speaks to the dangers of aesthetics, commodification and human desire.

Offstage, and as the nation's already wealthy landowners became even wealthier, 'imports of silk doubled in the 1590s'.[41] Like Moffett's *The Silkewormes and their Flies*, both *The Comedy of Errors* and *The Merchant of Venice* respond to the decade's uptick in the fabric's demand. When James VI/I acceded the throne in 1603, he took a far more proactive approach to silk production than Elizabeth I. Adopting the sericultural activities of other European royal families such as the Medici and the Valois, James spent the first decade of his reign attempting to advance the nation's economy by promoting domestic silk production.[42] The considerable silk-working skills of the Huguenot refugees who fled to London during the French wars of religion in the 1580s meant that labour wasn't a problem:[43] England merely needed to grow mulberry trees with which to feed the silkworms. After Henri IV's initiative, 'James ordered "those of ability" to distribute 10,000 mulberry plants at 3 farthings a plant or 6s, a hundred'.[44] The full titles of contemporaneous publications such as Nicholas Geffe's 1607 translation of Olivier de Serres' *The perfect use of silkewormes, and their benefit* and William Stallenge's

Instructions for the increasing of mulberie trees, and the breeding of silke-wormes, for the making of silke in this kingdome (1609) showed support for the monarch's plantational push. It's said by some that even Shakespeare planted a mulberry tree.[45]

At first glance, stories about James 'appoint[ing] a special governor of the chamber whose duties included carrying a couple of the wriggling insects "withsoever his majesty went"'[46] might suggest some kind of concern for the creatures themselves. Borlik explains that 'Many early moderns regarded all silkworms with...reverence, given the care lavished on them in sericulture'.[47] Indeed, Geffe/Seeres' observation that 'many beasts and strange plants, consent to live amongst us with requisite care'[48] and Stallenge's point that 'it is a principall part of that Christian care which appertaineth to Sovereigntie, to endeavour by all meanes possible...to beget as to increase among their people the knowledge and practice of all Artes and Trades' might include mindfulness for the silkworms.[49] This is not the case. The King's motives are driven by a 'project' of 'conspicuous display at court'[50] while the titles of Geffe/Serres' and Stallenge's books show that the 'use' and 'benefit' of Jacobean sericulture are designed to increase silk production for material gain and by extractive means.

We already know that Moffett's 1599 poem is interested in the silkworm's lifecycle, thus he doesn't shy away from talking about 'Their manner of dying'.[51] Silkworm death, according to Moffett, may be caused by a range of natural causes (including 'The smel of onyons, leekes, garlick, and new wheat,/ Shrill sounds of trumpets, drums or cleaving woode').[52] The poem also considers the silkworm's figurative demise as it transforms to join 'flying things'.[53] However, *The Silkewormes and their Flies* doesn't get to the real cost for sericulture's non-human actors. For the silkworm's protein fibre to be spun into thread, the living silkworm must be removed from the cocoon. Accordingly, Geffe/Serres' text shows 'The means to kill the Butterflies in the coddes'.[54] Near the end of the book, a closing woodcut depicts 'the fashion of the Engine, how to wind off the silke from the cods, with the furnaces and cawtherns [cauldrons] for that purpose'.[55] Stallenge instructs: 'You must either presently winde, or kil the wormes ... by laying the saide bottoms two or three daies in the Sunne, or in some Oven after the bread baked therein is taken out, and the fiercenesse of heat is alaied'.[56] The conclusion to the silkworm's lifecycle doesn't end in aerial liberation; it ends in their death by either boiling or baking.

Apart from the cocoon recalled in Nick Bottom's name, the Shakespearean plays I've discussed so far overlook the creaturely aspects of silk production in the style of Thomas Middleton's *The Revenger's Tragedy* (1607), which demands 'Does the Silkworm expend her yellow labours for thee?'.[57] Nonetheless, at roughly the same time that King James VI/I was trying to establish an English silk economy, *Othello* (another play set amidst the mercantile economy of early modern Venice) gives a prominent role to Moffett's 'little creepers'. At the tragedy's midpoint, Othello makes the 'ocular proof' (3.3.365) of Desdemona's supposed infidelity apparent:

> That handkerchief
> Did an Egyptian to my mother give.
> She was a charmer, and could almost read
> The thoughts of people. She told her, while she kept it
> 'Twould make her amiable, and subdue my father
> Entirely to her love; but if she lost it,
> Or made a gift of it, my father's eye

> Should hold her loathèd and his spirits should hunt
> After new fancies. She, dying, gave it me,
> And bid me, when my fate would have me wived,
> To give it her. I did so, and take heed on't.
> Make it a darling, like your precious eye.
> To lose't or give't away were such perdition
> As nothing else could match.
> ...
> There's magic in the web of it.
> A sibyl, that had numbered in the world
> The sun to course two hundred compasses
> In her prophetic fury sewed the work.
> The worms were hallowed that did breed the silk.
> And it was dyed in mummy, which the skilful
> Conserved of maidens' hearts. (3.4.53–73)

The audience is aware that Emilia has given the handkerchief, Desdemona's 'first remembrance from the Moor' (3.3.295), to Iago. In the wake of Othello's speech at 3.4.55–73, the object's loss leads to Desdemona's murder and shows *Othello*'s Venetian culture to be a perditious place on earth. As William C. Wyckoff observes, 'The mystery connected with the production of silk is used to advantage by Shakespeare. The fatal handkerchief whose loss brought about the death of Desdemona, was a silken fabric'[58] from 'hallowed' worms. Like Moffett's earlier poem, and as the silkworms take their place alongside the sibyl who 'sewed the work' and the artisans who made the dye 'Conserved of maidens' hearts', Shakespeare's play pulls contemporary sericultural processes into conversation with creative writing. Instead of domesticating georgic tropes like Moffett, Shakespeare turns Cinthio's

Gil *Hecatommithi* prose source into a complex metatheatrical moment bound up with fabric, fabrication and intertextual reproduction in which we glimpse the silkworms' part in constructing 'That handkerchief'.

My discussion started with a review of silk's protective qualities against the Covid-19 pandemic. Othello's account of his handkerchief's maternal legacy foregrounds the cloth's talismanic qualities. In a striking contrast to interpretations that view the handkerchief as white and largely representative of Desdemona's tragedy, Ian Smith's important essay 'Othello's Black Handkerchief' scrutinises the Moor's 'first remembrance' and shows that, 'dyed in mummy', the material is not white: it is black. Smith's analysis thus examines 'the handkerchief, its relation to Othello, ... its role in constructing an idea of blackness and race that places severe constraints on black subjectivity'[59] and ultimately interrogates:

> what reading whiteness in the handkerchief, despite evidence to the contrary, reveals about the habits and intellectual reflexes that inform our critical imagination. The essay asks whether this inveterate predisposition to see only a white handkerchief functions as an index to inherited critical frameworks that continue to circulate and shape the work of reading and producing knowledge in the field of early modern studies.[60]

Smith's carefully researched and closely argued discussion of silk production as set out in *Othello* 3.4.53–73 lays bare how 'Shakespeare takes the reference to the handkerchief's "Moorish fashion"' in Cinthio's *Gli Hecatommithi* 'more seriously than has been allowed, translating the racial and ethnic designation of the source in the presentation of a black handkerchief in the play'.[61] After reading Smith's meticulous argument, it's no longer possible to see *Othello*'s handkerchief as a piece of white silk embroidered with red strawberries. In Smith's words:

> Without a mention of strawberries in the source, the handkerchief is branded as foreign and unique in its exceptional Moorish design...Shakespeare's deliberate return to textile...produces an astute perception concerning the construction of blacks within a brutal materialist discourse. The black man as chattel, a nonhuman thing with the legal status of movable property on a colonial plantation estate, is familiar to modern audiences and readers as an historical image with a consequential legacy in the era of late capitalism.[62]

Smith's article provides a vital counterpoint to a version of Shakespeare Studies that relentlessly—despite evidence to the contrary—perpetuates white-fashioned worlds. I'll have more to say about this problematic desire to uphold *The Great White Bard* (to cite the title of Farah Karim-Cooper's essential book)[63] in the next section. While it's not his essay's focus of course, Smith's judicious unpacking of the silk handkerchief's genealogy and its significance for understanding Othello's role in the play simultaneously draws attention to the extractive treatment of living beings like 'hallowed worms' and Eurocentric attempts to appropriate silk production at large.

Inspired by European attainment, James continued to work on his ultimately doomed enterprise (it turns out that 'The

King was wrongly advised'[64] about the kinds of mulberry trees to plant). Trailing China's long-standing sericultural dominance from the Song dynasty (960–1279 CE) forwards,[65] other parts of the world developed domestic silk production in the seventeenth century with much more success. After placing restrictions on silk thread imported from China in 1685, for example, Japan's long-standing sericultural activities experienced 'a resurgence in domestic raw silk production'.[66] Cue Branagh's *As You Like It*.

At the start of this book, I suggested that *As You Like It* was Shakespeare's most surface-aware play. Coincidentally, the play's date of 1599 is contemporaneous with Moffett's *The Silkewormes, and their Flies*. In critical hindsight—that is with the sericultural backdrop I've provided above in mind—a silk-related network of ideas seems apparent in Branagh's cinematic version of the play. Indeed, this Lionsgate/HBO Films/The Shakespeare Film Company adaptation contextualises Shakespeare's Elizabethan comedy in a silken-surface-savvy way. Before the Shakespearean plot begins, Branagh adds three frames which tells the audience that:

> In the latter part of the 19th century, Japan opened up for trade with the West. Merchant adventurers arrived from all over the world, many of them English.
> Some traded in silk and rice and lived in enclaves around the 'treaty ports.'
> They brought their families and their followers and created private mini-empires where they tried to embrace this extraordinary culture, its beauties and its dangers…
> A dream of Japan
> Love and nature in disguise
> All the world's a stage. (00:44–01:47)[67]

With the first two frames implicitly setting the film in the Meiji era (1868–1912), a period of Japanese economic and social change,⁶⁸ the third frame dissolves into a brown and russet (silk?) curtain, which glides to the right to reveal (in a sort of discovery space) a small-scale Kabuki-inspired performance.⁶⁹ The opening shots juxtapose the intricately patterned silk of the on-stage kimonos with the block-colour nineteenth-century European dresses worn by two women spectators (who turn out to be Rosalind and Celia when Shakespeare's play eventually begins) (01:48–06:45): an apt prefatory comparison for a Shakespearean comedy that makes much of its heroines' knowledge of clothing as sites of social difference, which reaches a climax in the Forest of Arden with Rosalind-disguised-as-Ganymede-disguised-as-Rosalind (3.2.269ff).

The first six minutes of Branagh's *As You Like It* takes the form of the director's induction showing the overthrow of Duke Senior by his brother Duke Frederick. Another set of opening titles fashioned in an anglicised shodo-like format inform us that this part of the film (that is, the actual Shakespearean bit) is now sponsored by 'HBO/In association with BBC Films/ A Shakespeare Film Company Production'. Some film critics didn't know what to do with Branagh's concept. *Empire*'s two-star 2007 review remarked 'we're still not sure why Branagh set it in Japan'.⁷⁰ Since the film's original release date scholars have had more ideas about Branagh's 'dream of Japan'. Alexa Huang, for instance, considers Branagh's film in terms of 'a boomerang business', a 'twenty-first century phenomenon that is fueled simultaneously by globalized economic and cultural developments'.⁷¹ For Huang, 'Plays that have been traveling the world since his lifetime are now returning to Britain with many different hats' and Branagh's 'intercultural borrowing' might be viewed in both negative and positive ways.⁷² On the one hand, Branagh's *As You Like It* upholds the

West's Orientalist views of Japan; on the other hand, the film adaptation shows positive aspects of global trade. But what does the film tell us about silk? Some 14 years after the film's release, and following the break with the European Union in January 2020, 'on 11 September 2020 the UK and Japan [agreed] an historic free trade agreement…the UK's first major trade deal as an independent trading nation'.[73] Though Japan can't be seen as a main supplier to the UK, silk remains one of the noteworthy, imported materials.[74] Viewed now, Branagh's induction serves as shimmering celebration of sericultural trade. Though it clearly uses the fabric's visual aesthetics to draw the audience in (and here, the film director is more Herrick than the playwright), Branagh's *As You Like It* is in keeping with Shakespeare's socio-economic responses to silk in *A Midsummer Night's Dream*, *The Comedy of Errors*, *The Merchant of Venice* and *The Winter's Tale*. In its acknowledgement of the creature's worth for the handkerchief's production, *Othello* alone remains a noteworthy outlier while keeping quiet about the silkworm's wretched fate.

This section started by thinking about silk as a remarkable fibre for human protection from the Covid-19 virus. However, it's clear that humankind has scant compassion for the non-human beings 'that did breed the silk'. According to a BBC article from November 2022, we have made these extraordinary organisms into 'a strange mutant':

> For the vast majority of these hungry little caterpillars, their journey ends abruptly—and tragically—when their cocoons are tipped into hot water during the first step of silk processing. The developing moths inside are boiled alive and their cosy shelters become their tombs…The captive silk moths are so coddled, they've entirely lost the ability to

fly. During their week-long adult lives, they have no way of escaping predators and can only seek out a mate on foot— they're entirely dependent on humans to place males and females near one another.[75]

I agree with Borlik that Shakespeare's works remind us of the silkworm's labour, but it requires our critical effort to understand the full extent of humankind's pursuit of extractive industry no matter what the cost. In the end, I find myself paraphrasing Shakespeare's contemporary 'Does the Silkworm expend' their 'labours for thee?' and answering—without hesitation and with remorse—a resounding 'yes'.

NOTES

1 Max Berlinger, 'Fashion in the time of Covid-19: Luxury masks made of silk de chine, cotton poplin go mainstream', *The Economic Times*, 14 April 2020. https://economictimes.indiatimes.com/magazines/panache/fashion-in-the-time-of-covid-19-luxury-masks-made-of-silk-de-chine-cotton-poplin-go-mainstream/articleshow/75139536.cms
2 Michael Miller, 'Silk layer improves function of surgical masks', *University of Cincinnati News*, 24 May 2022. www.uc.edu/news/articles/2022/05/doubling-up-with-silk-improves-effectiveness-of-surgical-masks.html para 13, para 3.
3 BBC News, 'Covid: Disposable masks pose pollutants risk, study finds', 4 May 2021. www.bbc.co.uk/news/uk-wales-56972074 para 1.
4 Dagmar Schäffer, Giorgio Riello and Luca Molà, 'Introduction', in *Threads of Desire: Silk in the Pre-Modern World*, edited by Dagmar Schäffer, Giorgio Riello and Luca Molà (Suffolk: Boydell, 2018), pp. 1–18, p. 1.
5 Schäffer, Riello and Molà, 'Introduction', p. 1.
6 Todd A. Borlik, 'Shakespeare's mulberry: eco-materialism and "living on"', *The Shakespearean International Yearbook*, 15 (2015): 123–45, pp. 129–30.
7 Borlik, 'Shakespeare's mulberry', p. 124, p. 131.
8 Borlik, 'Shakespeare's mulberry', p. 131.
9 Borlik, 'Shakespeare's mulberry', p. 123.

10 Borlik, 'Shakespeare's mulberry', p. 128, p. 131.
11 Robert Herrick, 'Upon Julia's clothes', in *Hesperides, or, The Works both Human and Divine of Robert Herrick* (1648), X2r–X2v.
12 Hester Lees-Jeffries reminds us that '"Silk" is an umbrella term for many different kinds of textile'. 'Going in silks: some textile contexts for portraits of Charles I and Henrietta Maria', *Beyond the Label: Fitzwilliam College*, Cambridge (2019). https://beyondthelabel.fitzmuseum.cam.ac.uk/labels/going-in-silks para 2.
13 Anne E. Witte, 'Bottom's tangled web', *Cahiers Elizabethans*, 56.1 (1999): 25–39, p. 31.
14 Thomas Moffett, *The Silkewormes, and their Flies: Lively described in verse, by T. M. a country farmar, and an apprentice in Physic. For the great benefit and enriching of England* (London 1599). For excellent analyses of Moffett's poem, see further: Katharine A. Craik, 'These Almost Thingles Things': Thomas Moffat's The Silkewormes, and English Renaissance Georgic', *Cahiers Elizabethans*, 60:1 (2001): 53–66; Peter Auger, 'The natural history of *The Silkewormes, and Their Flies*', *Cahiers Elizabethans* 78:1 (2010): 39–45.
15 Moffett, *The Silkewormes, and their Flies*, n.p.
16 Moffett, *The Silkewormes, and their Flies*, B1r.
17 Moffett, *The Silkewormes, and their Flies*, n.p.
18 According to Victor Holliston, Moffett's verse is 'the first Virginian georgic poem in English, and contributed to the attempt, later encouraged by James I, to establish sericulture in England'. Victor Holliston, 'Moffett [Moufet, Muffet], Thomas [T.M.] (1553–1604, ODNB, para 7.
19 Moffett, *The Silkewormes, and their Flies*, B1r.
20 Moffett, *The Silkewormes, and their Flies*, K2r. Cited in Auger, 'The natural history of *The Silkewormes*', p. 40.
21 Moffett, *The Silkewormes, and their Flies*, K3v.
22 Auger, 'The natural history of *The Silkewormes, and their Flies*', 44.
23 Moffett, *The Silkewormes, and their Flies*, B1v.
24 Moffett, *The Silkewormes, and their Flies*, B2r.
25 Jacques Derrida, *The Animal That Therefore I Am*, translated by David Wills (New York: Fordham University Press, 2008), p. 5.
26 Moffett, *The Silkewormes, and their Flies*, C1r.

27 See, for example, Raphael Lyne, *Ovid's Changing Worlds: English Metamorphoses 1567–1632* (Oxford: Oxford University Press, 2001) and my own *Ovid and the Cultural Politics of Translation in Early Modern England* (London: Routledge, 2006).
28 Moffett, *The Silkewormes, and their Flies*, D1v.
29 Natale Conti, *Mythologiae* [1568], translated and annotated by John Mulryan and Steven Brown (Arizona: Arizona Centre for Medieval and Renaissance Studies, 2006), p. 885.
30 24 b. 'The cocoon of a silkworm'.
31 Borlik, 'Shakespeare's mulberry', p. 130.
32 For example, *An act for continuance of the statutes of perjury, for making of jails, for pewterers, and for sowing of flax and hemp* (1536–7). Note of acts', in *Journal of the House of Lords: Volume 1, 1509–1577* (London, 1767–1830), p. 102. British History Online http://www.british-history.ac.uk/lords-jrnl/vol1/p102
33 Wilfrid Hooper, 'Tudor sumptuary laws', *The English Historical Review*, 30: 119 (1915): 433–49, p. 433.
34 Teacher Pack, *William Shakespeare: Twelfth Night*, The Royal Shakespeare Company (2017). https://media.bloomsbury.com/rep/files/rsc-twelfth-night-teacherpack-2017.pdf p. 6.
35 Hooper, 'Tudor sumptuary laws', p. 433.
36 The British Library, 'Proclamation against excess of apparel by Queen Elizabeth', *Discovering Literature: Shakespeare and Renaissance*. www.bl.uk/collection-items/proclamation-against-excess-of-apparel-by-queen-elizabeth-i para 2.
37 Hooper, 'Tudor Sumptuary Laws', p. 437.
38 Cited in Hooper, 'Tudor sumptuary laws', p. 440.
39 *The Norton Anthology*, p. 719.
40 Ann Rosalind Jones, 'Cesare Vercelli, Venetian writer and art-book cosmopolitan', in *A Companion to the Global Renaissance: Literature and Culture in the Era of Expansion, 1500–1700*, edited by Jyotsna G. Singh, second edition (London: Wiley Blackwell: 2021), pp. 341–59, p. 341.
41 Linda Levy Peck, 'Creating a silk industry in seventeenth-century England', *Shakespeare Studies*, 28 (2002): 225–8, p. 225.
42 Peck, 'Creating a silk industry in seventeenth-century England', p. 226.
43 Peck, 'Creating a silk industry in seventeenth-century England', p. 225.
44 Peck, 'Creating a silk industry in seventeenth-century England', p. 225.

45 Borlik, 'Shakespeare's mulberry', p. 123.
46 Peck, 'Creating a silk industry in seventeenth-century England', p. 227.
47 Borlik, 'Shakespeare's mulberry, p. 131.
48 Nicholas Geffe, *The perfect use of silkworms, and their benefit* (1607), B3r.
49 William Stallenge, *Instructions for the increasing of mulberie trees, and the breeding of silkewormes, for the making of Silke in this Kingdome* (1609), A3r.
50 Peck, 'Creating a silk industry in seventeenth-century England', p. 227.
51 Moffett, *The Silkwormes, and their Flies*, Table.
52 Moffett, *The Silkwormes, and their Flies*, I2r.
53 Moffett, *The Silkwormes, and their Flies*, I3r.
54 Geffe, *The perfect use of silkworms, and their benefit*, L4r
55 Geffe, *The perfect use of silkworms, and their benefit*, O1r.
56 Stallenge, *Instructions for the increasing of mulberie trees, and the breeding of silkewormes, for the making of Silke in this Kingdome*, C3r.
57 3.5. 71–2 in Brian Gibbon (ed), *The Revenger's Tragedy*, second edition (London: Bloomsbury, 1991).
58 [William] C. Wyckiff, 'The Romance of a Caterpillar, Pt. II' [January 24, 1880] reprinted in *The American Entomologist*, 51.2 (2005): 80–1, p. 81.
59 Ian Smith, 'Othello's black handkerchief', *Shakespeare Quarterly* 64.1 (2013): 1–25, pp. 1–3.
60 Smith, 'Othello's black handkerchief', p.3, p. 25.
61 Smith, 'Othello's black handkerchief', p. 15.
62 Smith, 'Othello's black handkerchief', p. 15, p. 24.
63 Farah Karim-Cooper, *The Great White Bard: Shakespeare, Race and the Future* (London: Oneworld, 2023).
64 Victoria Finlay, *Fabric: The Hidden History of the World* (London: Profile, 2021). p. 342.
65 Schäffer, Riello and Molà, 'Introduction', in *Threads of Desire: Silk in the Pre-Modern World*, p. 3. As its title suggests, this edited collection is essential reading for the topic.
66 Sandra Schaal, *Discovering Women's Voices: The Lives of Modern Japanese Silk Mill Workers in Their Own Words*, translated by Jim Smith (Leiden: Brill, 2022), p. 29. See also Fujita Kayoko, 'Changing silk culture in early modern Japan: on foreign trade and the development of "national" fashion from the sixteenth to the nineteenth century', in *Threads of*

Desire: Silk in the Pre-Modern World, edited by Dagmar Schäffer, Giorgio Riello and Luca Molà (Suffolk: Boydell, 2018), pp. 295–321.
67 Kenneth Branagh (dir), *As You Like It* (HBO/Lionsgate, 2006).
68 See Tristan Grunow, 'Japan's Meiji restoration', *Origins: Current Events in Historical Perspectives*, January 2023. https://origins.osu.edu/read/japans-meiji-restoration?language_content_entity=en
69 Mark Thornton Burnett, *Filming Shakespeare in the Global Marketplace* (Basingstoke: Palgrave Macmillan, 2007), p. 158.
70 Anna Smith, '*As You Like It* review', *Empire Online*, 31 August 2007. www.empireonline.com/movies/reviews/like-review/ para 7
71 Alexa Huang, 'Boomerang Shakespeare: Foreign Shakespeare in Britain', *The Cambridge Guide to the Worlds of Shakespeare: Volume 2*, edited by Bruce R. Smith (Cambridge: Cambridge University Press, 2016), pp. 1094–1101, p. 1094.
72 Huang, 'Boomerang Shakespeare: Foreign Shakespeare in Britain', p. 1094, p. 1096.
73 The Department of International Trade, 'UK and Japan agree historic free trade agreement', Gov.UK, 11 September 2020. www.gov.uk/government/news/uk-and-japan-agree-historic-free-trade-agreement para 1.
74 See, for example, 'United Kingdom imports from Japan', 2022, *Trading Economics*. https://tradingeconomics.com/united-kingdom/imports/japan
75 Zaria Gorvett, 'How humanity created "Sky Puppies"', BBC Future, 15 November 2022. www.bbc.com/future/article/20221111-how-humanity-is-changing-the-worlds-insects para 9, para 13.

Skin

Curating complexion[1]

Nine

Peter Brathwaite's response to the task posed by The Getty Museum on 25 March 2020 to its Covid-19-isolated Twitter (now called X) community—'choose your favorite artwork; find three things lying around your house; recreate the artwork with those items and share with us'[2]—led to the exhibition and book-length project *Rediscovering Black Portraiture*. With the aim of '[making] the Getty Museum Challenge a matter of representation', Brathwaite had a clear goal. He writes:

> As a Black British opera singer, I am used to being a rare breed—still…By re-creating as many Black figures as possible, using the stuff in my house and the camera of my iPhone7, I would commit to rescuing their voices from oblivion—and in doing so, challenge how art history has been told.[3]

Eleven of Brathwaite's artworks formed the *Visible Skin* exhibition which opened at the King's College London's Strand Campus on 10 September 2021 in the new street-facing public space of Strand Aldwych in Westminster.

Focusing on Renaissance art, *Visible Skin*'s website features Brathwaite's 'restag[ings]' alongside the portraits originally painted between the fifteenth and the early eighteenth

DOI: 10.4324/9780429326752-10

centuries, which 'testify to the presence and prominence of Black life in Renaissance Europe'[4]: Jan Jansz Mostaert, 'Portrait of an African Man' (Christophe le More?) (1525–1530); Hendrick Heerschop, 'The African King Caspar' (1654); Diego Velázquez, 'Kitchen Maid with the Supper at Emmaus' (1617–1618); Jan Davidsz de Heem, 'Still Life with Moor and Parrot' (1641); Anonymous, 'The Virgin of Guadalupe' (1745); Vittore Carpaccio, 'The Miracle of the Relic of the True Cross on the Rialto Bridge' (1494); Anonymous, 'Fragment of a Retable: Saint Maurice' (1517); Christoph Weiditz, 'An African Drummer' (1529); Anonymous, 'The Paston Treasure' (c.1665); Jaspar Beckx, 'Don Miguel de Castro. Emissary of Kongo' (1643); Anonymous, 'Black Artist Completing a Portrait of Maria Anna of Austria, Queen of Portugal' 1683–1754). *Visible Skin*'s online notes explain how:

> Peter's images bring to the fore the contemporary importance of historical presence—turning this collection from a meditation of the past to a powerful interrogation of its role in the present. Recovering the diversity of Renaissance Europe involves interrogating the narrative of the Renaissance itself.[5]

Like Imtiaz Habib's *Black Lives in the English Archives 1500–1677* (2007),[6] Brathwaite pushes back at the idea of a Renaissance figured predominantly as white. At the same time, *Visible Skin* compels me, a white woman, to consider my complicity in maintaining the UK's structurally racist ideology, an ideology which allows whiteness to be *invisible*.

In terms of Shakespeare, the Renaissance narrative challenged by *Visible Skin* has such a pervasive reach that when a character's skin colour isn't mentioned, the default description is white. That a lack of racial literacy[7] remains dominant in the 2020s is worrying. Stuart Hall's 1997 lecture 'Race the Floating Signifier' opened with the words 'I want, at what you might think a rather late stage in the game, to return to the question of what we might mean by saying that race is a discursive construct, that it is a sliding signifier'.[8] In other words, and what he already thought to be 'at a rather late stage', Hall revisits poststructuralism's awareness of the slipperiness of meaning in language and ideology's import for holding meaning in place. Accordingly, ideas of race are constructed in and through culture. In the same year as Hall's lecture, Richard Dyer's book *White: Essays on Race and Culture* (1997) argued that 'As long as race is something only applied to non-white peoples, as long as white people are not racially seen and named, they/we function as a human norm. Other people are raced, we are just people'.[9] Here, then, is the reason why Shakespearean characters are deemed white unless they are explicitly described otherwise. Whiteness is unseen. One year earlier, Kim F. Hall's *Things of Darkness: Economies of Race and Gender in Early Modern England* (1996) made it very clear from the outset of her important book that

> descriptions of dark and light, rather than being mere indications of Elizabethan beauty standards or markers of moral categories, became in the early modern period the conduit through which the English began to formulate the notions of 'self' and 'other' so well known in Anglo-American racial discourses.[10]

Nearly 30 years after the publication of Kim F. Hall's and Richard Dyer's books and Stuart Hall's lecture, the UK continues to debate race in ill-informed ways.

When UK educators in 2022, for example, tried to update university courses to engage with a more inclusive worldview and pedagogy, the popular press turned that aim into headlines like this one announcing 'Universities drop Chaucer and Shakespeare as "decolonisation" takes root. Many British universities have sought to liberate their courses from "white, Western and Eurocentric" knowledge'.[11] This isn't my understanding of decolonisation. University educators are trying to create inclusive spaces for discussing why 'white, Western and Eurocentric' worldviews are privileged above others and how to rethink those authors who have been co-opted into sustaining white privilege. And as Maxime Cervulle argues, 'it is no longer possible to understand whiteness as other than the cultural façade of structural racism'.[12] In teaching spaces and beyond, thinking about current critical and creative afterlives of Shakespeare's works can help '[recover] the diversity of Renaissance Europe' and work towards opening academia itself to diverse scholars. In a discussion of how 'white ownership of Shakespeare and early modern studies traverses entire generations', Harry R. McCarthy suggests that 'The role of white scholars…might begin with recognizing who, historically, has been granted "leave to speak" in Shakespeare studies, and examining why exactly that has been so'.[13] Ambereen Dadabhoy and Nedda Mehdizadeh's 2023 book *Anti-Racist Shakespeare* has as much to teach white educators like me about white privilege alongside its 'aim to create spaces where students are exposed to theories of racial power and are equipped to develop strategies for resistance to hegemonic racial regimes'.[14] Indeed, Dadabhoy and Mehdizadeh's work has a great deal to teach many fans of Shakespeare outside the domains of primary, tertiary and higher education. The racist reactions to the Black cast of the Royal Shakespeare Company's (RSC) 2022 production of *Much*

Ado About Nothing 'set in an Afro-futuristic world'[15] confirm that Shakespeare's skin is socially and culturally under scrutiny by theatregoers who are keen to preserve white supremacy.[16]

Throughout *Shakespeare on the Ecological Surface*, I've suggested that premodernity's humoralism might be useful for confronting several concerns brought into focus by the coincidence of Covid-19 and climate crisis. If Brathwaite's important revisions of premodern artworks show how skin—the body's surface—is used to fashion cultural and social selfhoods, humoral sensibilities trouble purely superficial readings of cutaneous imagery. Skin isn't merely the body's obvious overlay. Rather, it's a name given to one strata of a complex fusion of corporeal matter. 'The skin', as Steven Connor explains, 'always takes the body with it'.[17] At the same time, premodernity's view of skin's porosity extends its reach beyond the human body to the environment at large. Premodernity's humoral understanding of skin thus has more in common with the term 'complexion', a word which has a wider meaning than just 'the natural colour, texture, and appearance of the skin'.[18] Taking its cue from Romanic and medieval Latin, 'complexion', broadly meaning 'combination, connection, association' and more specifically 'a person's humoral qualities', links human temperament to the elements.[19]

Please note well at this point. Humoralist thinking allows me to conceptualise a world of non-human and human correspondence. But Shakespeare's works remind me that humoral theories' relational sensibilities *must not* be confused with social equality. In fact, Shakespearean humoral theories help show how whiteness is invented and its privilege asserted.

Late sixteenth–and early seventeenth-century humoralism's interest in the interconnected conditions of lived experiences show how messy human selfhoods are culturally managed by people—often white people—in positions of power. The lengthy title of Thomas Newton's 1576 translation of a Latin treatise by the Dutch physician Lemnius Levinas *The Touchstone of Complexions*, for instance, sets out *easie rules & ready tokens so that every one may perfectly try, and throughly know...the exacte state, habite, disposition, and constitution, of his owne Body outwardly: as also the inclinations, affections, motions, & desires of his mynd inwardly*. While *The Touchstone of Complexions'* Latin epigraph announces 'nosce te ipsum' (know thyself), the book is really a manual for cataloguing and ordering bodies and their allied characteristics. Dedicated to Sir William Brooke, an influential Elizabethan courtier, Newton's translation makes plain how air quality and region influence inhabitants. Here, the book's citation of a few lines from Lucan's *Civil Wars* are telling:

> Such as in th' East and scorching Clymes
>   are bredde: by course of kind,
> And Countries influence, meycockes [effeminate] soft
>   By daily proofe we finde.
> The North, that colde and frostie it,
>   Such weaklings none doth breede:
> The folkes there borne no warres can daunt:
>   of death they have no dread.
> ...
> These laddes dare venture life and lymme,
>   in manly Martiall trade.[20]

The poem's articulation of social and cultural differences in these so-called 'geo-humoralist'[21] ways show how premodern Europe's selfhoods are made through a jumbled set of criteria embracing race and gender.

As part of a book concerned with analysing the body 'outwardly' to understand the 'mynd inwardly', it's not surprising to learn a bit later on that 'colour sheweth what humours be in the body'.[22] Newton's *The Touchstone of Complexions* doesn't go as far as, say, George Best's account of Martin Frobisher's 1578 journey to look for the northwest passage. Best's book lays out its racist agenda in passages which say things like

> We also among us in *England*, have blacke *Moores*, *Ethiopians*, out of all partes of *Torrida Zona*, whiche after a small continuance, can wel endure the colde of our Countrey…I conclude that blacknesse procéedeth not of the hotenesse of the Clime, but…the infection of bloud.[23]

Along the way, Best tells his readers:

> I my selfe have séene an Ethiopian as blacke as a cole broughte into Englande, who taking a faire Englishe woman to Wife, begatte a Sonne in all respectes as blacke as the Father was, although England were his native Countrey, & an English woman his Mother: whereby it séemeth this blacknesse procéedeth rather of some naturall infection of that man, whiche was so strong, that neyther ye nature of the Clime, neyther the good complexion of the Mother concurring, coulde any thing alter, and therefore we can not impute it to the nature of ye Clime.[24]

With an eyewitness account to support his authoritative mode of address, and certainty is an important part of the text, Best holds up the 'good complexion' of the 'faire English woman' who was the child's mother against the father's, 'an Ethiopian as blacke as a cole'. Kim F. Hall's *Things of Darkness* provides a thorough discussion of the 'Ethiopian-English marriage printed in' Best's work and explains why (after Stuart Hall) the 'passage demonstrates the "binary system of representation [that] constantly marks and attempts to fix and naturalize the difference between belongingness and otherness" that is typical of racist discourse'.²⁵ Related images of 'belongingness' are found in more recent newspaper articles about well-known figures such as this one from 2021, which describes the then Duchess of Cambridge (now the UK's Queen-in-waiting) as a 'true English rose' who 'let[s] her natural beauty shine through with "peachy fresh" make-up', leaving her 'complexion looking the "healthiest it has ever looked"'.²⁶ While separated from Newton's and Best's accounts by about four hundred years or so, there are obvious continuities between texts produced in the reigns of both Elizabeth I and II.

Despite writers like Newton's and Best's efforts to tie human characteristics to a mix of anatomical and environmental 'easie rules'—and, make no mistake, there are many other writers before and after them who use similar strategies—there is nothing remotely organic or elemental about the organisation of humans via 'complexion'. In fact, it's clear to me that the word is part of England's 'racecraft'.²⁷ Ayanna Thompson tells us that 'race is not a biological, scientific, or genetic reality. Race is a fiction. I'll repeat: race is not a real thing. Nor is race a stable category that refers solely to skin color, somatic aspects, or phenotypes'.²⁸ Steven Connor's chapter-length discussion of 'complexion' in the *Book of Skin* (2004) says that 'Shakespeare's use of the word, of which there are more than forty examples, illustrate [its] transition' of meaning from 'temperament' to the 'designat [ion of] the skin'.²⁹ While the meaning of the word 'complexion' is various and variable, like Peter Brathwaite's artwork for *Visible Skin*, Shakespeare's plays dramatise the curation of skin colour: *Titus Andronicus*, *The Merchant of Venice* and *Othello* show all too clearly that black corporeal surfaces are not treated the same as white ones.

In *Titus Andronicus*, the imperial household's Nurse instructs Aaron 'a Moor' to kill the baby he fathered with her mistress, the hyper-white Goth Queen/ Roman Empress Tamora.[30] Described by the Nurse as 'A joyless, dismal, black, and sorrowful issue' and a 'babe, as loathsome as a toad/ Amongst the fair-faced breeders of our clime' (4.2.66–68), the child simultaneously exposes the mother's adultery and Rome's racism. Aaron's response 'is black so base a hue?' (4.2.71) is an evident example of Shakespeare's probing dramaturgy. The starkly racist worldviews of Shakespeare's earliest Roman tragedy give way to the more nuanced racism of Shakespeare's Venice.

The Prince of Morocco, Portia's fifth suitor in *The Merchant of Venice* (1.3.104), makes a striking self-reflective reference to 'complexion' when first encountering Belmont's heiress:

> Mislike me not for my complexion,
> The shadowed livery of the burnished sun,
> To whom I am a neighbour and near bred.
> Bring me the fairest creature northward born,
> Where Phoebus' fire scarce thaws the icicles,
> And let us make incision for your love
> To prove whose blood is the reddest, his or mine.
> I tell thee, lady, this aspect of mine
> Hath feared the valiant. By my love I swear,
> The best regarded virgins of our clime

Have loved it too. I would not change this hue
Except to steal your thought, my gentle queen. (2.1.1–12)

Ian Smith explains that 'At his first appearance in the play, Morocco displays his body for scrutiny and immediately submits an apology for his physical difference: "Mislike me not for my complexion/ The shadowed livery of the burnished sun, / To whom I am a neighbour and near bred"'.[31] Because 'the lott'ry of [her] destiny' (2.1.15) means that she doesn't get to choose her husband, Portia's direct response to the Prince of Morocco's eloquent explanation of how his 'visible skin' compares to the 'fairest creature northward born' is conveniently evasive. Once Morocco fails 'the lott'ry' and leaves Belmont, Portia's comment to her maid Nerissa is much more direct in its racist views: 'A gentle riddance. Draw the curtains, go. / Let all of his complexion choose me so' (2.7.78–79). For Smith, Morocco resorts to 'the skilful use of elegant language to effect the "performance of whiteness" in order to mitigate the resistant stain of his skin colour'.[32] In this way, and with reference to Franz Fanon's *Black Skin, White Masks* (1952) Smith argues, 'Like an early modern Fanonian figure, Morocco surrenders himself to what he intuits is the racist Venetian ideology embodied by Portia, and as a black man selects the linguistic white mask'.[33] In the later Venetian play, Iago's use of the term 'complexion' in conversation with Othello about his wife's Desdemona's fidelity is overtly charged with racist views:

to be bold with you,
Not to affect many proposèd matches
Of her own clime, complexion, and degree,

> Whereto we see in all things nature tends,
> Foh, one may smell in such a will most rank,
> Foul disproportions, thoughts unnatural!
> But pardon me, I do not in position
> Distinctly speak of her, though I may fear
> Her will, recoiling to her better judgment,
> May fall to match you with her country forms
> And happily repent. (3.3.233–43)

Following Iago's recent reminder to Othello that 'She did deceive her father, marrying you' (3.3.210), a line which also recalls the character's racist views in 1.1 when he describes the married couple as 'an old black ram... tupping [a] white ewe' (1.1.88–89), he now speaks of Desdemona's rejection of 'many proposed matches/ Of her own clime, complexion, and degree'. In 11 lines, Iago's rhetorical reach embraces the etymological range of premodern 'complexion', from skin colour to personality traits, and points to the embedded nature and insidious power of his racism.

I now want to amplify a point I made in this book's Introduction. As a white woman I have much to learn about the privileges I've enjoyed and continue to enjoy because of my whiteness. Recent work in the field of Critical Studies in Whiteness (CSW), which calls for 'explicitly integrated critiques that analyse whiteness as part of a broader racial formation, which is material, affective and discursive',[34] has helped inform my understanding of my positionality. My greatest debt, however, is to the scholars and thinkers who identify as Black, Indigenous and People of Colour (BIPOC), are experts in Premodern Critical Race Studies (PCRS) and use Critical Race Theory (CRT) as praxis. Specifically, I want to acknowledge the labour by Sunita Abraham, Brandi

Adams, Fabiha Askari, Dennis Britton, Vanessa Corredera, Ambereen Dadabhoy, Nandini Das, Ruben Espinosa, Imtiaz Habib, Kim F. Hall, Margo Hendricks, Geraldine Heng, Jonathan Hsy, Islam Issa, Farah Karim-Cooper, Wendy Lennon, Arthur Little Jr, Nedda Mehdizadeh, Natasha Magigi, Mary Rambaran-Olm, Anandi Rao, Francesca Royster, David Sterling Brown, Ian Smith, Preti Taneja and Ayanna Thompson. The anti-racist work of the following conference and collectives are also crucial for continuing to shape my understanding of my positionality: the British Shakespeare Association Conference in 2019 at Swansea University, UK; #RaceB4Race; #ShakeRace; Lancaster University's Decolonising Network; Shakespeare's Globe Anti-Racist Shakespeare. I'm grateful for all the ways these foregoing individuals and groups teach me about 'the weight of whiteness'[35] at work in Anglo-American Shakespeare Studies and beyond.

Ayanna Thompson's 2016 Introduction to The Arden Shakespeare's edition of *Othello*, which replaced E.A.J. Honigmann's (first published in 1997), marks a particular turning point in my grasp of Anglo-American Shakespeare Studies and race. Generally accepted as the series that has 'long set the gold standard in annotated, scholarly editions of Shakespeare's plays', the Arden Shakespeare's *Othello* revised by Thompson—'Scholar. Activist. Theatre Practitioner'[36] and founder of 'an alternative conference series to build professional community by and for scholars of color working on issues of race in premodern literature, history, and culture' called *RaceB4Race*[37]—starts with two important questions:

> How and where to begin? An introduction like this one serves not only to frame William Shakespeare's *Othello*

> but also to prioritize its themes, topics, and contexts. While there are differences of opinion about how best to frame *Romeo and Juliet* (should one foreground early modern concepts of love before contextualizing certain literary forms, like the sonnet), the debates are rarely heated or political, and unsurprisingly they rarely replicate themselves in the court of public opinion. For *Othello*, however, the debates get extremely heated and traverse the terrain between the academic and the public…So is *Othello* a play about race?[38]

By navigating past and present ideologies which shape literary criticism and theoretical approaches to Shakespeare's tragedy, Thompson's Introduction in its entirety answers these two questions. One of Thompson's key points is that

> The play is not an inanimate object that never changes. Instead *Othello* is a dynamic organism that is affected by every hand that touches it—from the actors who perform the role, the creative writers who rewrite the plot, to the scholars who contextualise its various, disparate and interconnected histories.[39]

Under the rubric of what could be called PCRS and/or CRT, approaches to texts that bring discourses of race, power and knowledge to the fore, Thompson suggests that Mythili Kaul's edited collection *Othello: New Essays by Black Writers* (1996) 'helped initiate new approaches to the play'.[40] Kaul's collection, Thompson states, 'made it clear that black readers, audience members, scholars and artists may be experiencing a different play from white ones'.[41] Fast forward nearly a quarter of a century after the publication of *Othello: New Essays* by Black Writers to Shakespeare Studies on the other side of the Atlantic Ocean.

In the second decade of 2000s, the kinds of questions Thompson considered in her important introduction seem under pressure in the UK. Events such as the UK's government backlash against CRT in October 2020—Black History Month in the UK and the one following the 'summer of unrest' and Black Lives Matter's 'fight for justice' in the wake of George Floyd's death on 25 May 2020[42]—speak to its desire to control this aspect of my education and the education of others. Towards the end of the House of Commons' five or so hours debate on that year's Black History Month, for instance, the UK's then Minister of State for Equalities Kemi Badenoch stated:

> I want to speak about a dangerous trend in race relations that has come far too close to home in my life, which is the promotion of critical race theory, an ideology that sees my blackness as victimhood and their whiteness as oppression. I want to be absolutely clear that the Government stand unequivocally against critical race theory…any school that teaches….elements of critical race theory as fact…is breaking the law.[43]

Though Badenoch's speech bypassed tertiary education, it's hard to think that this earlier 'stand… against critical race theory' didn't influence the then Minister of State for Higher and Further Education Michelle Donelen's letter to universities on 27 June 2022 questioning their anti-racist initiatives such The Race Equality Charter.[44] To anyone who knows anything about CRT and the closely related PCRS, Badenoch's 2020 speech reduced the field's nuanced work to media-friendly soundbites. In a response to the debate, Kube Shand-Baptiste explained CRT's aims more closely:

> In education specifically, it would, in straightforward terms, mean teaching about the impact of structural racism, as well as incorporating a wider variety of texts on colonialism, its effects and its legacy into the curriculum.[45]

One of CRT's key concepts is 'intersectionality', a term coined by Kimberlé Crenshaw at the end of the 1980s to help explain how social privilege depends on a range of enmeshed social factors alongside race such as gender, class and sexuality.[46] It's not simply the case, as Badenoch's 2020 statement argues, that CRT upholds the binary logic of 'blackness as victimhood' and 'whiteness as oppression'. Structural racism is supported by an intricate network of what are often invisible agencies of power. As we've seen so far, structural racism's insidious impact can be shown very well via the scholarly and public reception of Shakespeare's plays and poems. Shakespeare Studies can show 'racecraft' in action.

I am not exempt from curating complexion and treating whiteness as the norm when studying Shakespeare's 'other race plays',[47] that is plays apart from those that feature Othello, Aaron, The Prince of Morocco and Shylock, the Jewish moneylender in *The Merchant of Venice*. When I now appraise my co-edited collection of essays *Twelfth Night: A Critical Reader* which was published in 2014,[48] for instance, I can see that the volume is avowedly white in terms of the editors, authors and the essays' outlooks. My co-editor and I have asked the publishers if we can revise the volume to openly address the collection's white privilege. At the time of writing, the publishers have sent the book out for external review to help with those revisions. We wait.

For now, and as I think about this current section on skin, I'm struck that *Twelfth Night* is noteworthy for its use of the word 'complexion' in three successive scenes of Act 2. In each episode, the term is used in the context of heteronormative sexual desire but with contrasting effects. In 2.3, Maria comes up with the plot to convince Malvolio that Olivia is enamoured with his 'complexion' (141) (among other things). It's evident she's been working away on the deception when the steward enters in 2.5 and remarks that 'Maria once told me that [Olivia] did affect me, and I have heard herself come thus near, that should she fancy it should be one of my complexion' (20–22). In the middle of these two scenes, Viola-as-Cesario and Orsino engage in an intimate conversation about the Duke's unrequited love for Olivia. Orsino asks about the kind of woman Viola-as-Cesario is in love with and they answer 'Of your complexion' (2.4.25). In the context of *Twelfth Night's* plot, complexion is a shorthand for showing the extent of the characters' understanding of their social standing. Though temporarily disguised as a lower-class youth, the aristocratic Viola knows her rightful place and ultimately resumes that place as Orsino's wife at the play's conclusion. By contrast, Malvolio is punished for exercising his rightful authority as the manager of Olivia's household: he is first tricked and then tortured by the play-text's group comprising Maria, another household labourer Fabian, Feste the Clown and Olivia's uncle Toby for wanting to elevate his role from his mistress' servant to her lover. Malvolio's line in Act 3 'I am not of your element' (3.4.112) underscores the play's humoral context at large and the character's lack of fit with the play's social ecology. Malvolio's only option is to exit the play at 5.1.365, announcing 'I'll be revenged on the whole pack of you', a sentiment which draws attention to *Twelfth Night's* violent vibe.

The racist responses to the RSC's production of *Much Ado About Nothing* I mentioned above show how visual aspects of plays in performance engage with contemporary cultural climates in ways that written texts simply can't. Just about a mile away from the opening of Brathwaite's *Visible Skin*, on the other side of the river Thames Shakespeare's Globe's Summer 2021 multicultural production of *Twelfth Night* was in the final month of its run. To coincide with the play's season and as part of Shakespeare's Globe's Anti-Racist Shakespeare series of free online talks designed to 'bring together scholars and artists of colour from a wide variety of backgrounds to examine Shakespeare's plays through the lens of race and social justice',[49] actor Natasha Magigi and academic Arthur Little Jr in conversation with Will Tosh considered *Twelfth Night*'s use of epithets like 'fair' to describe Olivia (1.2.30) alongside the numerous references to 'dark' (3.4.121; 4.2.36; 4.2.40) and 'darkness' (4.2.27; 4.2.50; 4.2.84; 5.1..293) associated with Malvolio's torment.[50] The play's dispersal of this imagery, Little explains, 'create[s] a racialized landscape that never quite materializes and/or it seems to me to be almost falling apart'.[51] If Malvolio's and Viola's association with the word 'complexion' in Act 2 scenes 2–5 contribute to the play's wider interests in social mobility, Magigi, Little and Tosh foreground some important ways diverse casting simultaneously magnifies *Twelfth Night*'s interest in 'service and status...around being Illyrian or not Illyrian'[52] and the importance of pluralistic representation.

Tosh and Magigi recall that in Shakespeare's Globe 2016 production (which featured Magigi as Maria)

> Viola's entry into Illyria was in a shipping container....and that spoke so vividly to modern migration...It was the time

of...channel crossings and of course things have got much worse. But you know that line 'What country friends, is this? [1.2.1]...She's on the the seashore. She's just washed up. It's...kind of vivid.⁵³

Magigi's own experience in the roles of Maria and Feste provides important insights into *Twelfth Night*'s treatment of skin, surface and race. 'Sometimes it is forgotten', Magigi says, 'that your audience sees whoever you cast...as that role' and regarding height, stature and skin colour 'You cannot pretend to be other than you are'.⁵⁴ In performance, actors' complexions count.

To reflect further on the invisibility of whiteness, structural racism and my own complicity in curating complexion I want to turn to a different medium to discuss two multicultural film versions of *Twelfth Night* which were released on the pay-to-view streaming platform Amazon Prime in 2018. Both adaptations—one produced by the American production company Hollywood Shakespeare⁵⁵ and the other by the UK's Shanty Productions⁵⁶—are filmed in the style of Joss Whedon's 2012 small-budget *Much Ado About Nothing*. Like Whedon, the films use everyday surroundings to render *Twelfth Night* in contemporary domestic settings. Before the pandemic, as Shakespearean adaptations released in what Tim Marshall calls *The Age of Walls*,⁵⁷ I understood the films as respective engagements with Trump's America and Brexit Britain and I originally viewed this pair of *Twelfth Nights* as comparative sites of social justice with particular takes on the racialised landscapes they depicted.

With a nod to the production's geographical location of California, Hollywood Shakespeare's Illyria is a glowing sun-filled island gently lapped by easy-going waves reminiscent of the Pacific Ocean's encounter with America's West Coast. Orsino, played by the African American actor Johnny Ramey, is a record producer; Amy Rapp's pale-skinned blonde Olivia is an heiress who lives in a house with an outside pool and a small phalanx of white staff. Malvolio is Olivia's publicist complete with a cell-phone earpiece in every scene (including one in the bathroom). Feste is an ageing musician. One of the most interesting aspects of this adaptation is its elevation of Fabian (Jessica Record) from a marginal character to a second fool in some kind of amatory relationship with Feste who sets up the film's main concept. As indicated by the telegraph poles and wires in the background of the film's opening shot of the littoral landscape and Malvolio's ever-present attachment to his cell phone, Illyria's inhabitants are deeply invested in social media and MMS messaging and Fabian is no exception. Before the main title, the film establishes its adaptive premise by way of Fabian typing up a celebrity blog entry which leads with news about Orsino and his would-be love interest Olivia. And so, Hollywood Shakespeare's *Twelfth Night* begins.

The film's US internet release date was what we now consider to be Twelfth Night itself (5 January 2018), a timely Christian festive event that underscores the production website's description of the plot as 'madcap mischief along the way'. This highly edited adaptation of Shakespeare's interrogative play—its subtitle is 'What you will', remember—seems keen to emphasise that 'madcap mischief' at the expense of *Twelfth Night*'s concerns for power, punishment and revenge. For instance, Hollywood Shakespeare's *Twelfth Night* generally achieves its brisk running time of 1 hour 18 minutes by cutting Malvolio's persecution in 4.2. The steward's last lines are an amalgamation of those at 3.4. 82–83 and 3.4.111–12, 'I discard you. Let me enjoy my private.

You are shallow things' (57:20). Sir Toby's response 'We'll have him in a dark room and bound' (3.4.121; 57:38) is Hollywood Shakespeare's only trace of the play-text's disturbing treatment of Malvolio in 4.2. Once heard—and it's very easy to miss—Sir Toby's quickly uttered line is a startling moment in the film. While Hollywood Shakespeare's *Twelfth Night* includes some non-verbal scenes which flag up Sir Toby's grief at the loss of his nephew (moments which ask the audience to reflect on his immediate disruptive behaviour with some empathy) his easily missed plan for Malvolio casts a disquieting shadow over Olivia's household. Sir Toby's instruction extends Maria's plan to humiliate her senior colleague much further: 'Madcap mischief' segues into punishment. Unlike Shakespeare's *Twelfth Night*, which dramatises the punishment of a man who tried to elevate his social standing and allows the audience to see his exit calling for revenge, Hollywood Shakespeare leaves me wondering what happened to Malvolio and what to make of the violence embedded in this white household.

By contrast with Hollywood Shakespeare's edits, Shanty Productions presents *Twelfth Night* in full. When I first saw the film (screened as an extracurricular activity for our undergraduate students in one of my university's lecture theatres in the first week of its release), I was struck by the inclusive view of England the production conveyed. For example, Ugandan British actor Sheila Atim plays the twins Viola and Sebastian among a multicultural cast. But now I'm not so sure about this adaptation's representation of diversity. Reviewing the film in the wake of May 2020, October 2020 and the disturbing news reports about the UK government's actions toward migrants and refugees in 2022, Shanty Productions' *Twelfth Night* is discomforting.

In a scene set up to bring Channel-crossing migrants to mind, Alex Downey as a kindly white sailor approaching Atim's shipwrecked Viola on the pebble-strewn shore is clearly out-of-whack with what the UK in 2022 is currently proposing in terms of refugees and asylum seekers.[58] Even without that specific context, his scene has much in common with Teju Cole's concept of a 'white saviour industrial complex', which is 'not about justice. It is about having a big emotional experience that validates privilege'.[59] This validation is upheld by Atim's Viola saying (as in the play):

> There is a fair behaviour in thee...
> And though that nature with a beauteous wall
> Doth oft close in pollution, yet of thee
> I will believe thou hast a mind that suits
> With this thy fair and outward character. (2.2.43–47)

While Viola's reading of the corporeal surface suggests that 'beauteous walls' can 'close in pollution', Shanty Productions' film endorses the white sailor's embodiment of 'a mind that suits/ With this thy fair and outward character'. I can see now how this scene facilitates an 'emotional response that validates privilege' and how I'm co-opted in its display of a 'white saviour industrial complex'.

These two filmed versions of Twelfth Night come out of differently difficult political situations, which, in their own ways, are ongoing. Yet it's noteworthy that Hollywood Shakespeare and Shanty Productions both cast white actors as Feste (Charles Baker; David Nellist), a figure who moves between Illyria's social spheres with ease. At the same time, these American-Anglo Twelfth Nights feature actors of colour as Antonio (Iseluleko Ma'at El 0; Zackary Momoh), a man who 'cannot without danger walk [Illyria's] streets' for fear of being captured by Orsino to account for a 'sea-fight' (3.3.25–6). As Tosh points out, Antonio 'isn't Illyrian' and 'doesn't get a happy ending at all'.[60] When Antonio and Orsino meet in *Twelfth Night*'s closing scene, the Count declares, 'That face of his I do remember well,/ Yet when I saw it last it was besmeared/ As black as Vulcan in the smoke of war' (5.1.45–47). When Orsino addresses him as a 'Notable pirate' and 'a salt-water thief' (5.1.63), Antonio refutes the accusations but confirms he is 'Orsino's enemy' (5.1.67–70). For Little, Orsino's revelation of Antonio's whiteness (which 'for the plot is sort of irrelevant') is an example of 'the sense of racialising going on in the play', that 'something is festering', but it's not explicit.[61]

Twelfth Night's opacity around race and identity in general allows the interrogation of what and how the human body means and who gets to construct that meaning. Shakespeare's play thus helps us to understand that complexion's vocabulary is not natural; that skins and bodies are not curated equally. In terms of Shakespeare and the invention of whiteness, these 2018 film adaptations from either side of the Atlantic engage with *Twelfth Night*'s examination of complexion even as they draw upon legacies of Anglo-American racism and embedded white privilege that multicultural casting cannot erase.

NOTES

1 I would like to thank Wendy Wall and Ruben Espinosa for commenting on earlier drafts of this section. I am very grateful for their expertise and time.
2 @GettyMuseum, 25 March 2020. https://twitter.com/GettyMuseum/status/1242845952974544896.
3 Peter Brathwaite, *Rediscovering Black Portraiture* (Los Angeles, CA: Getty Publications, 2023), pp. 9–10.
4 'Visible Skin', *Renaissance Skin* https://renaissanceskin.ac.uk/visibleskin/about/ para 5.

5 'Visible Skin', *Renaissance Skin* https://renaissanceskin.ac.uk/visibleskin/about/ para 12.
6 Imtiaz Habib, *Black Lives in the English Archives 1500–1677: Imprints of the Invisible* (London: Routledge, 2007).
7 For an informed summary of the phrase 'racial literacy', see Ambereen Dadabhoy and Nedda Mehdizadeh, *Anti-Racist Shakespeare* (Cambridge: Cambridge University Press, 2023), pp. 5–16.
8 A transcript of the lecture is published as Stuart Hall, 'Race, the floating signifier: what more is there to say about race', in Stuart Hall, *Selected Writings on Race and Difference*, edited by Paul Gilroy and Ruth Wilson Gilmore (Durham, NC: Duke University Press, 2021), pp. 359–73, p. 359.
9 Richard Dyer, *White: Twentieth Anniversary Edition* (London: Routledge, 2017), p. 1.
10 Kim F. Hall, *Things of Darkness: Economies of Race and Gender in Early Modern England* (New York: Cornell University Press, 1996), p. 2.
11 Craig Simpson, *The Telegraph*, 27 August 2022. www.telegraph.co.uk/authors/c/cp-ct/craig-simpson/.
12 Maxime Cervulle, 'Looking into the light: whiteness, racism and regimes of representation', translated by Emily F. Henderson, in Dyer, *White: Twentieth Anniversary Edition*, pp. xiii–xxxii, p. xvii.
13 Harry R. McCarthy, 'Leave to speak: white scholars, "Allyship", and Shakespeare Studies', *Shakespeare*, 17:1 (2021): 134–42, p. 136.
14 Dadabhoy and Mehdizadeh, *Anti-Racist Shakespeare*, p. 1.
15 BBC News, 'Royal Shakespeare Company: Director saddened by racist reaction to cast', 19 January 2022 www.bbc.co.uk/news/uk-england-coventry-warwickshire-60061769 para 5.
16 I follow Dadabhoy and Mehdizadeh's use of the term after Charles W. Mills: "white supremacy as a system, or set of systems, clearly comes into existence through European expansionism and the imposition of European rule through settlement and colonialism on aboriginal and imported slave populations". Dadabhoy and Mehdizadeh, *Anti-Racist Shakespeare*, pp. 10–12, p. 10.
17 Steven Connor, *The Book of Skin* (London: Reaktion, 2004), p. 29.
18 OED 4a.
19 OED complexion n. etymology, I.2.a. See also Connor, *The Book of Skin*, p. 19.

20 Thomas Newton (tr), *The Touchstone of Complexions* (1576), p. 13.
21 The term is Mary Floyd-Wilson's. See Floyd-Wilson, *English Ethnicity and Race in Early Modern Drama* (Cambridge: Cambridge University Press, 2003), p. 2.
22 Newton (tr), *The Touchstone of Complexions*, p. 89.
23 George Best, *A true discourse of the late voyages of discoverie, for the finding of a passage to Cathaya, by the northwest, under the conduct of Martin Frobisher generall devided into three books* (1578), p. 20, pp. 31–2.
24 Best, *A true discourse of the late voyages of discoverie*, p. 29.
25 Hall, *Things of Darkness*, p. 11.
26 Monica Greep, 'Kate Middleton looked "younger and more stylish" at the Royal Variety Performance thanks to her "casual, curly hairstyle" and fresh "English rose" complexion, experts reveal', *Mail Online*, 19 November 2021. www.dailymail.co.uk/femail/article-10220447/Kate-Middletons-complexion-healthiest-looked-peachy-fresh-make-up.html para 3.
27 Cited in Thompson who explains that 'racecraft' was 'coined by Karen E. Fields and Barbara J. Fields' to define how 'race is constructed by a social process'. Ayanna Thompson, 'Did the Concept of Race Exists for Shakespeare and His Contemporaries?: An Introduction', in *The Cambridge Companion to Shakespeare and Race*, edited by Ayanna Thompson (Cambridge: Cambridge University Press, 2021), pp. 1–16, p. 7.
28 Thompson (ed), 'Did the Concept of Race Exists for Shakespeare and His Contemporaries?: An Introduction', p. 7.
29 Connor, *The Book of Skin*, p. 20.
30 See Francesca Royster, '"White lim'd walls: whiteness and gothic extremism in Shakespeare's *Titus Andronicus*"', *Shakespeare Quarterly*, 52.4 (Winter 2000): 432–55, p. 432.
31 Ian Smith, 'The textile black body: race and shadowed livery' in *The Merchant of Venice*', in *The Oxford Handbook of Shakespeare and Embodiment*, edited by Valerie Traub (Oxford: Oxford University Press, 2016), pp. 170–85, p. 177.
32 Smith, 'The textile black body', p. 178.
33 Smith, 'The textile black body', p. 178.
34 Shona Hunter and Christi van der Westhuizen, 'Preface', in *The Routledge Handbook of Critical Studies in Whiteness*, edited by Shona Hunter and Christi van der Westhuizen (London: Routledge, 2021), pp. xx–xxvi, p. xxi.

35 See further Eduardo Bonilla-Silva, 'The invisible weight of whiteness: the racial grammar of everyday life in America', *Michigan Sociological Review*, 26 (2012): 1–15. See also Ruben Espinosa, *Shakespeare on the Shades of Racism* (London: Routledge, 2021), pp. 20–21.
36 Ayanna Thompson. https://www.ayannathompson.com/.
37 Ayanna Thompson. https://www.ayannathompson.com/raceb4race.
38 Ayanna Thompson. 'Introduction', in *The Arden Shakespeare Othello*, edited by E.A.J. Honigman (London: Bloomsbury, 2016), pp. 1–116, pp. 1–2.
39 Thompson, 'Introduction', pp. 4–5.
40 Thompson, 'Introduction', pp. 62–3.
41 Thompson, 'Introduction', p. 63.
42 *ITV News*, 29 December 2020. www.itv.com/news/2020-12-29/black-lives-matter-2020s-summer-of-unrest-and-the-fight-for-justice.
43 Hansard HC vol 682 cols 1011–12. https://hansard.parliament.uk/commons/2020-10-20/debates/5B0E393E-8778-4973-B318-C17797DFBB22/BlackHistoryMonth.
44 Michelle Donelan, Letter to Vice-Chancellors, WONKHE, 27 June 2022. chrome-extension://efaidnbmnnnibpcajpcglclefindmkaj/https://wonkhe.com/wp-content/wonkhe-uploads/2022/06/Letter-Regarding-Free-Speech-and-External-Assurance-Schemes-1.pdf.
45 Kuba Shand-Baptiste, 'The government has no intention of taking racism seriously', *The Independent*, 22 October 2020. www.independent.co.uk/voices/critical-race-theory-racism-kemi-badenoch-black-history-month-bame-discrimination-b1227367.html para 6.
46 Kimberlé Crenshaw, 'Demarginalizing the intersection of race and sex: a black feminist critique of antidiscrimination doctrine, feminist theory and antiracist politics', *University of Chicago Legal Forum*, 1 (1989): 139–67.
47 I'm particularly indebted here to the work of David Sterling Brown. See, for example, David Sterling Brown and Jennifer Lynn Stoever, 'The sound of whiteness or, teaching Shakespeare's "other race plays" in five acts', *Folger Critical Race Conversations*, 16 July 2020. www.youtube.com/watch?v=iBGSh4h-74U&t=960s.
48 Alison Findlay and Liz Oakley-Brown (eds), *Twelfth Night: A Critical Reader* (London: Bloomsbury, 2014).
49 *Twelfth Night/ Shakespeare and Race*, Shakespeare's Globe, 7 October 2021. www.youtube.com/watch?v=lg1R46guBeY.

50 David Sterling Brown observes that 'Twelfth Night's Fool asserts, in an otherwise nonsensical reply to Sir Andrew, 'My Lady has a white hand' (2.3.27). Despite its randomness, his statement denotes how the lady's white hand is a recognisable racialised symbol whose meaning, which is linked to domestic relations and relationships, transcends class boundaries'. David Sterling Brown, 'Shake thou to look on't: Shakespearean white hands', in White People in Shakespeare: Essays on Race, Culture and the Elite (London: Bloomsbury 2023), pp. 105–19, p.107.
51 Twelfth Night / Shakespeare and Race, 15:21–15:35.
52 Twelfth Night / Shakespeare and Race, 25:42.
53 Twelfth Night / Shakespeare and Race, 38:30–38:48.
54 Twelfth Night / Shakespeare and Race, 38:30–38:48.
55 Twelfth Night, directed by Ned Record (2018). www.hollywoodshakespeare.com/films.html.
56 Twelfth Night, directed by Adam Smethurst (2018). https://shantyproductions.com/.
57 Tim Marshall, Divided: Why We're Living in an Age of Walls, revised edition (London: Elliott and Thompson, 2018).
58 For example, 'Rwanda threat raised UK asylum seekers' suicide risk, clinicians said', The Guardian, 1 September 2022. www.theguardian.com/uk-news/2022/sep/01/rwanda-threat-raised-uk-asylum-seekers-suicide-risk-clinicians-said.
59 Teju Cole, Twitter (now X.com), 8 March 2012. https://twitter.com/tejucole/status/177810262223626241?s=20? See also Teju Cole, 'The white–savior industrial complex', The Atlantic, 21 March 2012. www.theatlantic.com/international/archive/2012/03/the-white-savior-industrial-complex/254843/?utm_source=copy-link&utm_medium=social&utm_campaign=share.
60 Twelfth Night / Shakespeare and Race, 33:32.
61 Twelfth Night / Shakespeare and Race, 35:23–36:21.

Stage

Disposable globes

Ten

Philip Breen's Covid-19-delayed production of *The Comedy of Errors* for Stratford-Upon-Avon's Royal Shakespeare Company (RSC) opened on 13 July 2021 at the Lydia and Manfred Gorvy Garden Theatre, 'a temporary outdoor performance space in the Swan Gardens, with safe outdoor seating for up to 500 people'.[1] For those of us who felt too anxious to watch a play in less well-ventilated areas, the Garden Theatre's airy structure assembled on a tree-lined greenspace next to the river Avon provided reassurance. My various attempts to see that summer's production of *Hamlet* at the Theatre Royal Windsor (the one with the then 82-year-old Ian McKellan in the title role) were thwarted by my Covid-19-related apprehensions. By contrast, sitting in a shaded back-row seat on a sunny September afternoon felt joyful, a site-specific happiness that worked well with the kind of play about to be performed upfront. Set in Ephesus—a bustling ancient Greek coastal city—this Stratfordian September afternoon chimed with this production's design and *The Comedy of Errors*' upbeat tempo. The sea-blue patina of the stage paralleled the sky-blue sky aloft (or was it the other way around?) while the red polypropylene tiers of seats drew attention to the playing space's cooler hues. With a knowing nod to our current difficulties navigating indoor and outdoor spaces, Breen's *The Comedy of*

DOI: 10.4324/9780429326752-11

Errors 'integrated use of surgical masks and hand gel'.² Rather than ignoring Covid-19, this three-month Garden Theatre production (it ran to 26 September 2021) made a creative virtue from a deadly virus and worked hard to engage its surface-aware audience in specific ways.

Three months before the production opened, Greg Doran explained that 'the outdoor theatre would be a confidence-building bridge'.³ It's likely that Shakespeare's plays were originally conceived with professional amphitheatres in mind (a point discussed in the earlier parts of this book). This kind of playing space exploited humoral sensibilities to their utmost; amphitheatres embrace the audience rather than alienate it. And there was something especially inclusive about the time, space and place of the RSC's project:

> To be comfortable is to be so at ease with one's environment that it is difficult to distinguish where one's body ends and the world begins. One fits, and by fitting, the surfaces of bodies disappear from view. The disappearance of the surface is instructive: in feelings of comfort, bodies extend into spaces, and spaces extend into bodies.⁴

For the first time since the pandemic took hold, I was doing something I'd not done in over three years: watching Shakespeare onstage, not on a screen. Unlike *The Comedy of Errors*' Antipholus of Syracuse who wonders 'Am I in earth, in heaven, or in hell?' (2.2.212), seated in the space between the indivisible 'world of the play' and the 'world of the place',⁵ I was at ease.

If the RSC repurposed *The Comedy of Errors* for Covid-19-related circumstances in summer 2021, the company also had a wider stance on re-purposeful action in general. *Shakespeare on the Ecological Surface*'s opening section discussed the actor-vists who stormed the RSC during the 2012 production of *The Tempest* to protest the company's BP sponsorship. In the same year, the RSC began to 'measure and report [their] carbon emissions, covering energy and water consumption, waste and travel by staff' on an annual basis.[6] 'Our monitoring', the RSC's website says, 'is improving all the time, and we now capture more data than ever before, which can be used to make further environmental improvements'.[7] As the RSC's first production since lockdown, Breen's *The Comedy of Errors* was avowedly committed to this mission of repurposing in material terms. It's reassuring to learn that the Garden Theatre's 'stage and seating will go into storage for future use'.[8] A quick but necessary comment on those arresting red seats. A type of plastic, 'polypropylene is not biodegradable under normal circumstances' and its manufacture needs petroleum, natural gas, 'large amounts of water and power'.[9] Like any plastic, it's far from an ideal fabric for the environment. However, 'It is recyclable and can be reused if [polypropylene] products are properly recycled'.[10] Polypropylene is also strong and easily cleaned. All these factors are key for the RSC's combined aims for the Garden Theatre to function as a 'confidence-building bridge' in view of Covid-19 and 'environmental improvements'. In this final section, I want to work out from the RSC's and my own first post-lockdown theatrical experience to think further about Shakespeare, waste and the environment. First, I consider London's Elizabethan, Jacobean and twentieth-century Globe theatres before thinking about replica, reconstructed and pop-up globes. Specifically, I question whether the rise and fall of temporary Globe theatres are emblematic for privileged white Western approaches to the environment itself.

Though the RSC's initiative is a relatively new one, and Breen's 2021 production of *The Comedy of Errors* engages with the intersecting contemporary crises underpinning *Shakespeare on the Ecological Surfaces*, London's Shakespeare's Globe has a history bound up in another mode of recycling. In some respects, and as the story of the Globe illustrates, all theatres are temporary. As theatre historians comprehensively discuss, the first Globe which opened on London's Bankside in 1599 used 'reclaimed timbers of the Theatre in Shoreditch that was disassembled by its owners and removed to a newly leased site on Bankside in December 1598 and January 1599'.[11] There's a rich backstory about *The Battle That Gave Birth to the Globe*[12] involving early modern theatre impresarios, recalcitrant landlords, religious factions, class riots, law suits and aristocratic neighbours. Early modern environmental issues such as 'noise pollution' and 'waste' are also linked to the anti-theatrical diatribes in circulation,[13] and Randall Martin writes about 'the environmental crisis that prompted this thrifty recycling'.[14] Nonetheless, there's little contemporary evidence that this upcycling of the Theatre's expensive oak is for sustainable rather than economic reasons. With the much-cited reference to 'this wooden O' (Prologue 13), Chorus in *Henry V* reminds us that Shakespeare's writings are often self-reflexive about the matter and form of purpose-built theatres. And within playing spaces at large, the stage itself receives comment, most famously in Jaques' line 'All the world's a stage, / And all the men and women merely players' (*As You Like It* 2.7.138–39). *The Merchant of Venice*'s Antonio says something similar in that play's opening scene. 'I hold the world but as the world', he says, 'A stage where every man must play a part' (1.1.77–78). Drawing on 'the theatrum mundi, [an] early modern commonplace…

wherein the world is likened to the stage',[15] both Jaques and Antonio are alert to the kinds of human precarity built into the stage design of a theatre like the Globe. Organised into three vertical sections—hell (under the stage), earth (on the stage) and heaven (above the stage)—Shakespeare's stage is not a horizontal surface upon which the action merely stands. According to Tiffany Stern, though

> characters [such as Jaques and Antonio] are referring to early modern ideas about the structure of the universe, with heaven above, earth in the middle, and hell below, they are also all, simultaneously, making references to the structure of the theatres in which they perform.[16]

The tensions between humans, the environment they inhabit and textual space are also witnessed at the start of Sonnet 15:

> When I consider every thing that grows
> Holds in perfection but a little moment,
> That this huge stage presenteth nought but shows
> Whereon the stars in secret influence comment;
> When I perceive that men as plants increase,
> Cheerèd and checked even by the selfsame sky;
> Vaunt in their youthful sap, at height decrease,
> And wear their brave state out of memory:
> Then the conceit of this inconstant stay
> Sets you most rich in youth before my sight,
> Where wasteful time debateth with decay,
> To change your day of youth to sullied night;
>   And all in war with time for love of you,
>   As he takes from you, I engraft you new.

The difference between characters like Jaques and Antonio and the sonneteer is that the verse throbs with a belief in the poem's ability to immortalise the addressee against 'decay'. Shakespeare's stage, then, is at once material, metaphysical and multilayered.

As we saw in section 'Two: Smoke', the 1599 Globe famously burnt down following the first performance of *Henry VIII: All Is True* on 29 June 1613 and Sir Henry Wotton's description of the amphitheatre's rapid conflagration shows the power of elemental agency above human effort. The decision to rebuild the second Globe (completed in 1614) with a tiled roof instead of the cheaper yet combustible thatch highlights the economy's role in sustainable living. Non-human and human preservation takes mindful long-term investment:

> Policymakers and politicians rely on economics to guide decision making on the risks posed to society by climate change. Central to that task are economic models. They can be used to estimate the future costs of climate change impacts such as heatwaves, flooding and sea-level rise, alongside the economic benefits of preparing for them. Models also estimate the costs and benefits of measures that reduce greenhouse gas (GHG) emissions.[17]

In the analytical spirit of sixteenth-century relational sensibilities, Shakespeare's Globe might be seen as a microcosmic example of twenty-first century Western approaches to climate emergency itself. Changes must be made but questions about financial cost govern the debate.[18]

In terms of the second Globe theatre (though the analogy might be prophetic for the earthly globe itself), the building's planned resilience against non-human intervention didn't matter: humans pulled it down. The Long Parliament's closure of all theatres in 1642 included the Globe; two years later, the English Civil Wars led to the theatre's '[rasure] by the Puritans'.[19] Opened in 1997, and drawing on the cultural legacies of the earlier playing spaces, the current Shakespeare's Globe celebrated its twenty-five-year anniversary in 2022. Like the RSC's goal to reduce its carbon footprint, the Globe 'established an Environmental Sustainability Taskforce, endeavouring to action the Theatre Green Book and commit to climate action' and 'made a commitment to progress towards net zero greenhouse gas emissions by 2050'.[20] With the overarching aim to 'celebrate Shakespeare's transformative impact on the world by conducting a radical theatrical experiment' via 'unique historic playing conditions',[21] it remains to be seen if the Globe's approach will have a 'transformative impact' on 2020's climate emergency. But it won't be for want of trying. On 22 April 2020—Earthday and about a month into the UK's first lockdown—the Globe Ensemble uploaded their performance of Titania's speech from *A Midsummer Night's Dream* 2.1.87–117 as 'Shakespeare's Letter to the Earth':

> But with thy brawls thou hast disturbed our sport.
> Therefore the winds, piping to us in vain,
> As in revenge have sucked up from the sea
> Contagious fogs which, falling in the land,
> Have every pelting river made so proud
> That they have overborne their continents.
> The ox hath therefore stretched his yoke in vain,
> The ploughman lost his sweat, and the green corn
> Hath rotted ere his youth attained a beard.
> The fold stands empty in the drownèd field,

> And crows are fatted with the murrain flock.
> The nine men's morris is filled up with mud,
> And the quaint mazes in the wanton green
> For lack of tread are undistinguishable.
> The human mortals want their winter cheer.
> No night is now with hymn or carol blessed:
> Therefore the moon, the governess of floods,
> Pale in her anger washes all the air,
> That rheumatic diseases do abound;
> And thorough this distemperature we see
> The seasons alter: hoary-headed frosts
> Fall in the fresh lap of the crimson rose,
> And on old Hiems' thin and icy crown
> An odorous chaplet of sweet summer buds
> Is, as in mock'ry set. The spring, the summer,
> The childing autumn, angry winter change
> Their wonted liveries, and the mazèd world,
> By their increase now knows not which is which;
> And this same progeny of evils comes
> From our debate, from our dissension.
> We are their parents and original.[22]

Part of the King and the Queen of the Fairies' quarrelling entrance into the play, Titania's retort to Oberon is 'A startling speech with analogies to today's new climate consciousness'.[23] Held together by alliterative words such as 'disturbed', 'disease', 'distemperature' and 'dissension'—synonyms for out-of-ordered states—the thirty-one lines of Titania's speech encapsulate seasonal strife so great that even 'the mazèd world…now knows not which is which'. Channelled as a call to environmental action, Titania's words foreground ecological discomfort.

As a production aware of the 'mazèd world' caused by the immediate effects of Covid-19 alongside the ongoing threat of the earth's environmental emergency, Breen's *The Comedy of Errors* asks its audience to consider disposable commodities. In this book's section 'Eight: Silk' I discussed this commonly called 'madcap comedy'[24] as one focused on material matters. Perhaps because so much pre-production attention was devoted to the Garden Theatre's installation (see the time-lapse video on You Tube showing its construction for instance)[25] the RSC's pro tem playing space brings *The Comedy of Errors'* interests in architectural surfaces to the fore. Bound in one day's action, Shakespeare's shortest play excels in making anthropocentric impulses a key aspect of its dramaturgy. Following Egeon's purposeful, storm-driven biographical account at the play's start, the ensuing fast-paced action takes place in and around inns called the Centaur (1.2.9, 1.2.104, 2.2.2, 2.2.9, 2.2.16, 4.4.144, 5.1.412), the Tiger (3.1.96) and the Porcupine (3.1.117, 3.2.165, 4.1.49, 5.1.223, 5.1.276), while Antipholus of Ephesus' home is known as the Phoenix (1.2.75, 1.2.89, 2.2.11). All these human-made structures are obviously named after real and imagined non-human creatures, a not-so-subtle reference to the ongoing determination of homo sapiens for hierarchical order.

Given that *The Comedy of Errors'* tragic potential works out from Egeon's illegal entry into Ephesus, it's striking that the plot's farce-like comedy relies on social status too: where you come from; who you are (or are not); who you know (or don't). If it seems a stretch to say that *The Comedy of Errors* asks its audience to interrogate human privilege at large, we only need to look at the play's treatment of its bondsmen. Patricia Akhimie points out that:

> Dromio of Syracuse and Dromio of Ephesus voice a knowing critique of the class system, describing the difference between servants or slaves and their masters in terms of race, as a somatically marked difference rather than a matter of fate...[they] draw attention to the fact that their frequent bruising (a somatic mark inscribed by means of the beating hands of social superiors) has less to do with their own acts that with their status as slaves/servants.[26]

To this physically brutal world, we can add Dromio of Syracuse's repulsive treatment of Nell, Adriana's kitchen-maid and Dromio of Ephesus' wife. In one of the play's key moments of mistaken identity, the Syracusan bondman runs onto the stage bemused by this unfamiliar woman who clearly thinks he is her husband. During 3.2, and in response to Antipholus of Syracuse's questions, Dromio of Syracuse produces what could be called a contra blazon of the kitchen-maid's body parts. Of her complexion, he says it's 'Swart like my shoe, but her face nothing like so clean kept. For why?—She sweats a man may go overshoes in the grime of it' (3.2.101–3). And her name, he reckons, functions as an aptronym: 'Nell...her name and three-quarters—that's an ell ['more than a yard'][27] and three-quarters—will not measure her from hip to hip' (3.2.108–10). Furthermore, 'She is spherical like a globe' and Dromio of Syracuse 'could find out countries in her' (3.2.113–14). As the men figuratively travel around this corporeal 'globe', Nell's body is marked and mapped via Ireland, Scotland, France, England, Spain, America, the Indies, Belgium and the Netherlands (3.2.115–36). The Syracusan dialogue dovetails with the textual/sexual manoeuvres of the probably contemporaneous poem by John Donne, Elegy 19 'To His

Mistress Going to Bed'. Here, the poem's narrator speaks of his lover's body as his 'America!', his 'Newfoundland' and his 'emperie' to conquer.[28] Unlike Donne's early verse, of course, Dromio of Syracuse's aim is denigration not consummation. The scene's ideological freight is brought to light in Jeffrey R. Wilson's reading:

> Nell remains off-stage as Dromio delivers a sardonic blazon that commingles the stigma of obesity with that of race, claiming to see ethnicities in her physical features: the bogs of Scotland in her buttocks, the 'heirlessness' of France in her baldness, the fire of Spain in her breath, the jewels of the Indies in her nose, but—importantly—no English whiteness in her teeth...Dromio's stigmatizing is colonial, in a manner of speaking, treating the self, white and thin, as normal and good, and the other, swart and fat, as inferior and therefore an object of degradation and ridicule.[29]

Wilson's deft analysis exposes the Syracusans' ability to make a disposable commodity out of the unseen kitchen-maid. This episode is thus yet another instance of how assumed positions of human power fashion the world and its inhabitants as they see fit. In 3.2.115–36 Nell's body isn't like a globe: it *is* the globe.

As we saw in section 'Three: Sky', sixteenth- and early-seventeenth-century Eurocentric culture enjoyed structuring the surfaces of terrestrial and celestial globes. It's easy to see how Dromio's response to Nell is in keeping with this contemporaneous orbicular discourse. The regeneration of London's Globe theatre from 1599 onwards is an extension of this project too. To paraphrase W.B. Worthen, in 'Reconstructing the Globe' we are 'constructing ourselves'.[30] And in a binary world view, a structuralist perspective that surfaces facilitate, selfhoods need others. If *The Comedy of Errors* raises questions about the politics of difference—from globes to the Globe—then what might a planetary proliferation of Shakespeare's Globes suggest?

Apart from the site-specific rejuvenations of the Elizabethan theatre in Southwark, replicas of Shakespeare's Globe have been built in other parts of the capital, England, Europe and East Asia, for example: London (Earls Court 1912); Stratford-upon-Avon, England (1964); Germany (Neuss 1991, Baden-Württemberg 2000); Rome (2003); North America (Chicago 1934), San Diego 1935, Texas 1966, Utah 1977; Tokyo (1988).[31] Yu Jin Ko explains:

> Not surprisingly, it has often been said that this phenomenon is the work of the same forces of globalization that have transformed the world's economies and cultural modes of activity and exchange…all surviving replicas outside London…exist exclusively in industrial countries.[32]

In approximately half a millennium, the rise in Shakespeare's Globes mirrors the rise in industrial globalisation itself. Along with the oil-slick streams I discussed in section 'One: Slick' it's not too difficult to imagine the earth's surface studded with multi-sided polygonal shapes: outcrops of Shakespeare's Globes all using the brand of the Bard to mark the earth's surface in celebration of a white, Western human creativity that has much in common with Dromio of Syracuse's colonisation of Nell. But if we zoomed back in to examine the history of Pop-Up Globes, a specific short-lived phase of replicas that burst on the scene between 2016 and 2021, we might learn a bit more about human vulnerability in the wake of material desires.

Pop-up Globes are connected to the rejuvenated and replica Globes I consider above. At the same time, these twenty-first century facsimiles are related to another kind of structural emulation which Marion O'Connor calls 'reconstructive Shakespeare', a mode best represented by William Poel's late-Victorian practice:

> For a production of *Measure for Measure* in November 1893, Poel set the stage of the Royalty Theatre with a simulacrum of an Elizabethan theatre... photographs taken of the set in use at least six different times show that, in all but minor decorative details, it had been modelled on the amphitheatrical Swan, of which a (second-hand) sketch, the so-called De Witt drawing, had surfaced only in 1888.[33]

Miles Gregory used the term Pop-Up Globe to name the venture he started in Auckland, New Zealand, in 2016 and which was originally designed as a one-off event to join Shakespeare 400 (the hashtag given to the many events commemorating Shakespeare's death in 1616). After undergoing a cultural metamorphosis caused by watching a performance of *Henry V* at London's Shakespeare's Globe in its opening year (he tells of how he started studying law at university and ended up with a PhD in Shakespeare and Performance),[34] Gregory's project is a mixture of personal and professional passion. Twelve months before the Pop-Up Globe's inaugural season, Gregory explained:

> Three years ago I moved back to my native New Zealand with my growing family. One Sunday I was reading a pop-up book to my oldest child, Nancy. Shakespeare's Globe popped up. 'Can we go there?', she asked. I explained that the Globe was a long way away—and then I stopped, and thought.

> A pop-up Globe.
>
> A full-scale, temporary, working theatre space that precisely replicates the dimensions of the Globe Theatre. That can quite literally 'pop-up' anywhere.
>
> Built from honest, simple materials—scaffolding, plywood, paint—using the smoke and mirrors of theatre to 'stage' the Globe so successfully that on first viewing you think you're looking at the real thing. A way for people who will never be able to travel to London to unlock the incredible power and beauty of Shakespeare performed in the space for which it was written.
>
> Because it's all about experience. And there's something about a pop-up theatre that captures the world-changing magic of Shakespeare's original.[35]

While it's easy to see Gregory's aim to use 'honest, simple materials' as part of an idealised recursive discourse, a nostalgic view of the past as a golden age reclaimed, there's something appealing about his description that speaks to the humoral qualities inscribed in Shakespearean texts and the audience's reception of them. Most of all, I'm arrested by Gregory's observation that 'To see this iconic building briefly juxtaposed with the skyline of your own city, far away from London, will be a tremendously exciting moment'.[36] In this line, Gregory describes the kinds of relational sensibilities I've been interested in throughout *Shakespeare on the Ecological Surface*: an interwoven connection of sky, city, self and Globe unhampered by surficial division. But if we revisit the 'honest, simple, materials' comprising the Pop-Up Globe, they're fabrics ushered into the commercial arena during the late-nineteenth and early twentieth-century industrial age: steel or aluminium; thin layers of wood; resin, pigments, binders, solvents and additives.[37] Underpinned by extraction (steel and aluminium) and felling (plywood), these destructive ecological transformations seem far removed from the 'world-changing magic' Gregory invokes.

Gregory's Pop-Up Globe initiative seemed to pave the way for temporary Shakespearean amphitheatres on either side of the Atlantic. In 2017, Angus Vail proposed The Container Globe, 'a low-cost alternative to brick-and-mortar theaters', which can be built 'anywhere there's containers and scaffolding—i.e. pretty much everywhere. Imagine a Detroit Globe, or an Africa Globe… a Wellington and a Mumbai Globe'.[38] Conceived as 'a Punk Reimagining of Shakespeare's Theatre'[39] with its first location slated for Detroit (the Michigan city renowned for Motown and automobiles), The Container Globe references twentieth-century pop culture while simultaneously promoting cost-effective practices. That same year, the city of York in England gained approval[40] for 'Europe's first-ever temporary Shakespearean theatre' to be erected.[41] Based on the design of Southwark's Rose Theatre (1587) rather than the Globe and made from 'state-of-the-art scaffolding, corrugated iron and timber',[42] the 13-sided playing space opened in June 2018.[43] A second short-term theatre opened at Blenheim Palace, Oxfordshire, in the summer of 2019.[44] Reviewing the trend she smartly termed 'Globe-al Dominance' in 2018, even though the fabric, 'the motives for the replicas, and their funding differs for each site',[45] Michelle Manning notices an obvious but salient coherence: 'it is the Globe's distinctive shape that is appearing across the world'.[46] Impelled by the same principles at work in the Syracusan twins' dialogue about Nell and the sonneteer's aim to halt human decay through writing, these arborescent protuberances speak to humankind's aspirations to make its mark. But to what end? What no one could foresee was the swift suppression of this fad by economic and/or environmental forces. Largely due to the fragility of the post-Brexit national economy, the UK's celebrated undertakings in York and Oxford ended after the 2019 season.[47] At the time of writing, The Container Globe remains a virtual reality. The pandemic ended Gregory's project in March 2020. In the context of climate crisis, it's hard to see these transitory theatres as anything other than examples of humankind's wider interests in mass production. To the capitalist refrain of 'fast fashion' and 'fast food', we might add 'disposable Globes'. The problem is inherent in the concept of globes in the first place.

In *Down To Earth: Politics in the New Climatic Regime*, (after Husserl) Bruno Latour draws attention to the globe as a 'Galilean object':

> The idea—the revolutionary idea—of grasping the earth as one planet among others, immersed in an infinite universe of essentially similar bodies, can be traced to the birth of modern sciences...The progress of this planetary vision has been enormous. It defines the cartographer's globe, the globe of the earliest earth sciences. It makes physics possible.[48]

The globe-as Galilean object—this exemplar of human innovation and privilege—also renders our planet as surface. And that's the problem. Because of globes, 'we have begun to see less and less of what is happening on Earth'.[49] Shakespeare's Globe is thus overtly connected to the 'Galilean object' rather than the interconnected components comprising the earth's climate system: atmosphere, hydrosphere, cryosphere, biosphere, pedosphere and humans. If we're serious about tackling climate change via Shakespeare, perhaps we could start by calling Shakespeare's Globe something else. With ecocritical thinkers like James Lovelock and Lyn Margulis, Gabriel Egan and Bruno Latour very much in mind, and if only to draw attention to the earth's networked ecology, I suggest Shakespeare's Gaia.[50]

NOTES

1. The Royal Shakespeare Company, Garden Theatre. www.rsc.org.uk/your-visit/our-theatres/garden-theatre para 1.
2. Mark Lawson, 'The Comedy of Errors review—glorious fun in the RSC's garden', The Guardian, 21 July 2021. www.theguardian.com/stage/2021/jul/21/the-comedy-of-errors-review-rsc-garden-theatre-stratford para 5.
3. Mark Brown, 'RSC plans Stratford Garden reopening', The Guardian, 22 April 2021. www.theguardian.com/stage/2021/apr/22/rsc-plans-stratford-garden-theatre-for-summer-reopening para 8.
4. Sara Ahmed, The Cultural Politics of Emotion, second edition (Edinburgh: Edinburgh University Press, 2014), p. 148.
5. I'm drawing here on Randall Martin's Social Sciences and Humanities Research Council of Canada-funded project Cymbeline in the Anthropocene, specifically 'Place'. www.cymbeline-anthropocene.com/shakespeare-ecocriticism-place para 1.
6. The Royal Shakespeare Company, 'Environmental Responsibility'. www.rsc.org.uk/about-us/environmental-responsibility para 1.
7. The Royal Shakespeare Company, 'Environmental Responsibility', para 1.
8. 'The Garden Theatre's deconstruction begins', The Royal Shakespeare Company. https://www.rsc.org.uk/news/garden-theatre-deconstruction-begins para 6.
9. 'Is Polypropylene Bad for the Environment?', Our Endangered World: Protect Earth. www.ourendangeredworld.com/eco/is-polypropylene-bad/ paras 8, 1–2.
10. Is polypropylene bad for for the environment?', para 8.
11. Gabriel Egan, 'The 1599 Globe and its modern replica: Virtual reality modelling of the archaeological and pictorial evidence', Early Modern Literary Studies, 13 (April 2004). https://extra.shu.ac.uk/emls/si-13/egan/index.htm para 1.

12 I'm referring here to the subtitle of Chris Laoutaris' absorbing book *Shakespeare and the Countess: The Battle That Gave Birth to the Globe* (London: Penguin, 2014).
13 Laoutaris, *Shakespeare and the Countess: The Battle That Gave Birth to the Globe*, p. 282.
14 Randall Martin, *Shakespeare and Ecology* (Oxford: Oxford University Press, 2015), p. 1
15 John Gillies, 'Globe/Theatrum Mundi', *The Cambridge Guide to the Worlds of Shakespeare*, edited by Bruce R. Smith (Cambridge: Cambridge University Press, 2016), pp. 60–99, p. 60.
16 Tiffany Stern '"This Wide and Universal Theatre": The Theatre as Prop in Shakespeare's Metadrama', in *Shakespeare's Theatre and the Effects of Performance*, edited by Farah Karim-Cooper and Tiffany Stern (London: Bloomsbury, 2013), pp. 11–32, p. 17.
17 Richard Black, 'Climate economics: Costs and Benefits', *The Energy and Climate Intelligence Unit*, 17 January 2022. https://eciu.net/analysis/briefings/climate-impacts/climate-economics-costs-and-benefits para 1.
18 According to W.B. Worthen, 'the [1997] Globe is the first thatched building in London since the great fire; the thatching is treated with retardant and the roof is protected from fire by a sprinkler system'. W.B. Worthen, 'Reconstructing the Globe: Constructing Ourselves', *Shakespeare Survey*, 52 (1999): 33–45, p. 39.
19 Tim Fitzpatrick, 'Reconstructing Shakespeare's second Globe using "computer aided design" (CAD) tools', *Early Modern Literary Studies*, Special Issue 13 (April 2004). https://extra.shu.ac.uk/emls/si-13/fitzpatrick/ para 1.
20 See 'Downloads: Annual Review 2021–2022', *Shakespeare's Globe*. www.shakespearesglobe.com/discover/about-us/#downloads p. 3.
21 'Downloads: Annual Review 2021–2022', *Shakespeare's Globe*, p. 1.
22 Alys Daroy, 'Shakespeare and climate change', *Shakespeare's Globe*, 22 April 2020. www.shakespearesglobe.com/discover/blogs-and-features/2020/04/22/shakespeare-and-climate-change/
23 Daroy, 'Shakespeare and climate change', para 8.
24 'The Comedy of Errors', *Shakespeare's Globe*. www.shakespearesglobe.com/whats-on/the-comedy-of-errors-2023/ para 1.

25 'Building the Lydia and Manfred Gorvy Garden Theatre', www.youtube.com/watch?v=OXSyZH8x53Q
26 Patricia Akhimie, 'Bruised with adversity: reading race in *The Comedy of Errors*', in *The Oxford Handbook of Shakespeare and Embodiment*, edited by Valerie Traub (Oxford: Oxford University Press, 2016), pp. 186–96, p. 188.
27 *The Norton Shakespeare*, p. 744n.
28 John Donne, 'To His Mistress Going To Bed', in *The Complete Poems of John Donne*, edited by Alexander B. Grosart (London: Robson and Sons, 1872), pp. 223–4, line 27, line 29. According to Achsah Guibbory 'Donne most likely wrote [the elegies] when he was a young man in his early twenties, attached to the Inns of Court. Elizabeth was queen of England and Petrarchan poetry was popular'. Achsah Guibbory, 'Erotic Poetry', in *The Cambridge Companion to John Donne*, edited by Achsah Guibbory (Cambridge: Cambridge University Press, 2006), pp. 133–48, p. 134.
29 Jeffrey R. Wilson, 'Nell's obesity', *Stigma in Shakespeare*. https://wilson.fas.harvard.edu/stigma-in-shakespeare/nell%E2%80%99s-obesity para 2.
30 Worthen, 'Reconstructing the Globe: constructing ourselves', title.
31 See Clara Calvo, 'Exhibiting the past: Globe replicas in Shakespearean exhibitions', *The Shakespeare International Yearbook* (2016): 65–86; Yu Jin Ko, 'Globe Theatre replicas', in *The Cambridge Guide to the Worlds of Shakespeare*, edited by Bruce R. Smith (Cambridge: Cambridge University Press, 2016), pp. 1084–94.
32 Yu Jin Ko, 'Globe Theatre replicas', p. 1084, p. 1091.
33 Marion O'Connor, 'Reconstructive Shakespeare: reproducing Elizabethan and Jacobean stages' in *The Cambridge Companion to Shakespeare on Stage*, edited by Stanley Wells and Sarah Stanton (Cambridge: Cambridge: Cambridge University Press, 2002), pp. 76–97, pp. 78–80.
34 Miles Gregory, 'Guest blog: Why I want to build a pop-up Globe Theatre', *WhatsOnStage*, 9 May 2015. www.whatsonstage.com/london-theatre/news/guest-blog-miles-gregory-pop-up-globe-theatre_37771.html/ paras 1–4.
35 Gregory, 'Guest blog: Why I want to build a pop-up Globe Theatre', paras 7–11.

36 Gregory, 'Guest blog: Why I want to build a pop-up Globe Theatre', para 13.
37 Daniel Palmer Jones and David Henry Jones inaugurated the use of steel in UK scaffolding in 1913 when they 'were commissioned to do much-need repair work to Buckingham palace'. *Business and Industry Connection Magazine*, 4 April 2018. www.bicmagazine.com/departme nts/lift-transport/april-2018-scaffolding-leader-celebrates-100-years-strong/ para 2; 'It was not until the 1850s that plywood started to be used on an industrial scale'. 'A short history of plywood in ten-ish objects', *Victoria and Albert Museum*, 15 July 2017 to 12 November 2017. https://www.vam.ac.uk/articles/a-history-of-plywood-in-ten-objects para 2; 'The explosion of ready mixed paint, the first truly modern paint, came with the industrial revolution', 'The History of Paint', *Building Materials*, 28 November 2022. www.buildingmaterials.co.uk/info-hub/painting-decorating/the-history-of-paint para 4.
38 The Container Globe. www.thecontainerglobe.com/ para 1.
39 Angus Vail, 'The Container Globe: a punk reimagining of Shakespeare's theatre', *TEDxJerseyCity*, 8 February 2016. www.youtube.com/watch?v=ZNryP2AP_CA
40 R. J. King, 'Shipping Container Replica of Shakespeare's Globe Theatre Plans Detroit Premier', *Business*, 13 May 2022. https://www.dbusiness.com/daily-news/shipping-container-replica-of-shakespeares-globe-theatre-plans-detroit-premiere/
41 'Ambitious plans unveiled for Yorkshire's answer to Shakespeare's Globe', *The Yorkshire Post*, 11 September 2017. www.yorkshirepost.co.uk/news/exclusive-ambitious-plans-unveiled-for-yorkshires-answer-to-shakespeares-globe-1772056 para 1.
42 Andrew White, 'Shakespeare's theatre pops up', *The Northern Echo*, 22 May 2019. www.thenorthernecho.co.uk/news/local/northyorkshire/17655778.shakespeares-theatre-pops/ para 3.
43 'Shakespeare cast unites for first rehearsals of pop-up plays in York', *The Yorkshire Post*, 30 April 2018. www.yorkshirepost.co.uk/news/shakespeare-cast-unites-for-first-rehearsals-of-pop-up-plays-in-york-296 987 paras 2–3.
44 'Shakespeare's Rose Theatre to pop up at Blenheim Palace', *Belfast Telegraph*, 22 April 2019. www.belfasttelegraph.co.uk/entertainment/news/shakespeares-rose-theatre-to-pop-up-at-blenheim-palace/38041173.html.

45 Michelle Manning, 'Globe-al dominance: the rise in reconstructed Globe theatres', *The Folger Shakespeare Library: Shakespeare and Beyond*, 27 March 2018. www.folger.edu/blogs/shakespeare-and-beyond/rise-in-reconstructed-globe-theatres/ para 4.
46 Manning, 'Globe-al dominance: the rise in reconstructed Globe theatres', para 9.
47 'Pop-up Shakespeare Rose theatre firm 'facing liquidation', *BBC News*, 25 September 2019. www.bbc.co.uk/news/uk-england-york-north-yorkshire-49828283
48 Bruno Latour, *Down To Earth: Politics in the New Climatic Regime*, translated by Catherine Porter (London: Polity, 2018), p. 67.
49 Latour, *Down To Earth: Politics in the New Climatic Regime*, p. 70.
50 See Ian Enting, 'Gaia theory: Is It Science Yet?', *The Conversation*, 12 February 2012 https://theconversation.com/gaia-theory-is-it-science-yet-4901 para 1; Gabriel Egan, 'Shakespeare and ecocriticism: The unexpected return of the Elizabethan World Picture', *Literature Compass*, 1 (2003): 1–13; Bruno Latour, *Facing Gaia: Eight Lectures On The New Climate Regime*, translated by Catherine Porter (Cambridge: Polity, 2017).

Afterword

Surface futures

I started this book as the world's humans donned masks for protection against the Covid-19 virus. As I write the Afterword on 8 June 2023, images of an amber-hewn New York dominate the news. The masks worn for protection against the Covid-19 virus are now worn as a shield against New York's smoke-filled air caused by raging Canadian fires. If Covid-19 drew attention to the human body's navigation of ecological surfaces, then the earth's climate emergency puts that singular experience into the planetary domain. Michel Foucault proposed that 'the body is the inscribed surface of events'.[1] But the body is just part of a wonderful, webbed network of 'vibrant matter'[2] that relies on each component to play its part (to almost cite *As You Like It* (2.7.141)).

Is this latest extraordinary elemental episode further proof that we've entered what Stephen J. Pyne calls the Pyrocene, 'an informing principle (in a literary sense) by which to understand the world our pact with fire has made'?[3] Humans, as Pyne's understanding of his own term as 'an informing principle' points out, like to organise geological time in neat eras like this one to add to the couple already mentioned in this book: Anthropocene and Myxocene. To these three geological phases, we could also add Symbiocene, Glenn A. Albrecht's proposition. Albrecht argues:

DOI: 10.4324/9780429326752-12

> I argue that the next era in human history should be The Symbiocene (from the Greek sumbiosis, or companionship). I created this concept in 2011 as an almost instinctive reaction against the very idea of the Anthropocene (Albrecht 2011). The scientific meaning of the word 'symbiosis' implies living together for mutual benefit and I wish to use this profoundly important concept as the basis for what I hope will be the next period of Earth history. As a core aspect of ecological and evolutionary thinking, symbiosis and its associated symbiogenesis, affirms the interconnectedness of life and all living things (Scofield and Margulis 2012).[4]

Isn't this what Shakespeare's plays and poems explore?

As humankind struggles to come up with new terms to define the superficial, that is the visible, effects of environmental crisis, the avoidance of a word in Shakespeare's writings encourages a reflection on 'the interconnectedness of life and all living things'. And as I've argued throughout *Shakespeare on the Ecological Surface*, the idea of surface is so bound up with depth—its apparent opposite—that it immediately suggests boundary, partition and detachment. Shakespeare's eschewal of the word 'surface' draws attention to this seemingly solid concept. The organisation of this book into ten sections headed by alliterative keywords—Slick, Smoke, Sky, Steam, Soil, Slime, Snail, Silk, Skin, Stage—brings a few surficial strands together to think about the environment at large.

There's an assemblage of surfaces I've depended on while writing and thinking about the three questions I asked at the start of *Shakespeare on the Ecological Surface*:

> What are the implications of the surfacing of the word 'surface' itself?

How does a consideration of Shakespearean surfaces help to explore premodern cultural politics?

To what extent does thinking about surfaces in Shakespeare's texts and their afterlives put a spotlight on twenty-first century ecological concerns?

Without the bound pages of type holding the plays and the poems together, obviously 'Shakespeare' as we know it would not exist. Indeed 'books and documents possess their own agential and material force, shaping us and our human world as much as we shape the written objects we consume and preserve'.[5] Shakespeare cannot solve humankind's ecological emergency. However, early modern Europe's interest in the elemental analogies between earth, air, fire and water that drives so much of Shakespeare's overtly surface-free writing is an outlook that twenty-first century Western culture would do well to critically consider. And then act. Our collective non-human and human existence depends on it.

NOTES

1 Michel Foucault. 'Nietzsche, genealogy, history', in *Language, Counter-Memory, Practice: Selected Essays and Interviews*, edited by Donald F. Bouchard (Ithaca, NY: Cornell University Press, 1977), p. 148.
2 I'm alluding to Jane Bennett, *Vibrant Matter: A Political Ecology of Things* (London: Duke University Press, 2010).
3 Stephen J. Pyne, 'The Pyrocene', www.stephenpyne.com/disc.htm para 1.
4 Glenn A. Albrect, 'Exiting the Anthropocene and entering the Symbiocene', *Psychoterratica*, 17 December 2017. https://glennaalbrecht.wordpress.com/2015/12/17/exiting-the-anthropocene-and-entering-the-symbiocene/ para 10.
5 Bruce Holsinger, *On Parchment: Animals, Archives, and the Making of Culture from Herodotus to the Digital Age* (New Haven, CT: Yale University Press, 2022), pp. 289–90.

Further reading

Everything I've acknowledged in *Shakespeare on the Ecological Surface* has helped shape my approach to the primary texts. However, there are four books that have informed my contribution to the Spotlight on Shakespeare series in immeasurable ways and which I recommend to anyone wanting to read, to think and to take ecological action via Shakespeare—and beyond.

Randall Martin's *Shakespeare and Ecology* (2015)[1] brings together a rich web of contextual knowledge, Shakespeare scholarship and environmental activism. Alongside Martin's writing, I also endorse the collaborative, interdisciplinary and intercontinental project led by Martin, *Cymbeline in the Anthropocene* (2020–22): 'a ground-breaking experiment in ecodramaturgy dedicated to representing global ecological relations through consciously situated and culturally differentiated stage adaptions of Shakespeare's late tragicomic play.'[2]

I've learned a great deal from Martin about ecological Shakespeare. For knowledge about ecotheoretical approaches to literature and the environment more broadly, three connected anthologies are essential reading. First published in 2013, Jeffrey Jerome Cohen's edited collection of sixteen essays *Prismatic Ecology: Ecotheory Beyond Green* extends extant ideas of what ecology looks like, from Bernd Herzogenrath's meditation 'White' to Timothy Morton's philosophically driven thought piece 'X-Ray', which probes the agency of nonhuman perception.[3] If the first volume in this triumvirate uses colour to shape its overarching argument, the second book takes up the topic that's closest to *Shakespeare on the Ecological Surface*. As its title suggests, Jeffrey Jerome Cohen's and Lowell Duckert's gathering of essays *Elemental Ecocriticism: Thinking with Earth, Air, Water, and Fire* (2015) is underpinned by Empedoclean views and the editors' 'Eleven Principles of the Elements'.[4] Fourteen writers respond to these principles including Anne Harris, 'Pyromena: *Fire's Doing*', Sharon O'Dair, 'Muddy Thinking', and Julian Yates, 'Wet?'[5] *Veer Ecology: A Companion for Environmental Thinking* (also edited by Jeffrey Jerome Cohen and Lowell Duckert, 2017) invites 30 authors to address the book's concept by choosing a titular verb as their critical springboard. Each verb launches a deliberately vital take on the 'whirled',[6] for example Cord J. Whitaker on 'Remember' and Teresa Shewry on 'Hope'.[7]

There's one book that was published too late for me to take close account of in *Shakespeare on the Ecological Surface* (and I'd already taken an excessive amount of time writing such a short monograph). In my Afterword, I briefly draw attention to the surfaces enfolded in the published versions of Shakespeare's works. I'm excited to take up Bruce Holsinger's ideas about 'ecocodicology', and how:

> the strange history of parchment teaches us [that] books and documents possess their own agential and material force, shaping us and our human world as much as we shape the written objects we consume and preserve.[8]

It seems to me that Shakespeare on the ecodicological surface is worth much more thought.

NOTES

1 Randall Martin, *Shakespeare and Ecology* (Oxford: Oxford University Press, 2015).
2 *Cymbeline in the Anthropocene*. www.cymbeline-anthropocene.com/outcomes-and-goals para 1.
3 Jeffrey Jerome Cohen (ed), *Prismatic Ecology: Ecotheory Beyond Green* (Minneapolis: University of Minnesota Press, 2013).
4 Jeffrey Jerome Cohen and Lowell Duckert (eds), *Elemental Ecocriticism: Thinking with Earth, Air, Water, and Fire* (Minneapolis: University of Minnesota Press, 2015).
5 Anne Harris, 'Pyromena: fire's doing', in Cohen and Duckert (eds), *Elemental Ecocriticism: Thinking with Earth, Air, Water, and Fire*, pp. 27–54; Sharon O'Dair, 'Muddy Thinking', in Cohen and Duckert (eds), *Elemental Ecocriticism: Thinking with Earth, Air, Water, and Fire*, pp. 134–57; Julian Yates, 'Wet?', in Cohen and Duckert (eds), *Elemental Ecocriticism: Thinking with Earth, Air, Water, and Fire*, pp. 183–208.
6 Jeffrey Jerome Cohen and Lowell Duckert (eds), *Veer Ecology: A Companion for Environmental Thinking* (Minneapolis: University of Minnesota Press, 2017), p. 3.
7 Cord J. Whitaker, 'Remember', in *Veer Ecology: A Companion for Environmental Thinking*, pp. 106–21; Teresa Shewry, 'Hope', in *Veer Ecology: A Companion for Environmental Thinking*, pp. 455–68.
8 Bruce Holsinger, *On Parchment: Animals, Archives, and the Making of Culture from Herodotus to the Digital Age* (New Haven, CT: Yale University Press, 2022), pp. 289–90.

Index

Note: Endnotes are indicated by the page number followed by "n" and the note number e.g., 119n16 refers to note 16 on page 119. The acronym "RSC" is used for the "Royal Shakespeare Company".

Acheloüs 35–6
active reflexivity 15–16, 23n38
activism: climate emergency 122; environmental 26, 39, 104, 236; of RSC 30–1, 38–9; social 2, 4
Adonis 45, 86, 88–90, 91, 92–3, 94, 95, 96, 98, 144
agency: elemental 217; of non-human perception 237; of the snail in *Thersites* 156; of surfaces and media 66
air 3, 6, 7, 17, 34, 85, 86, 191, 219, 233, 235; and smoke 52, 55, 56; and sky 65, 72, 73, 74, 77, 78, and soil 113, 114, 116
air travel 64
Akhimie, Patricia 220–1
Albertinus, Aegidius 145
Albrecht, Glenn A. 233–4
Alchemist, The 57–8
All Is True (*Henry VIII*) 51, 58, 217
allegory 71, 91, 104, 106, 117, 130
All's Well That Ends Well 79
Almagest 71, 79
Amato, Joseph A. 3–4
Anthropocene 5–6, 60, 122, 124, 233, 234, 236
Antipholus of Ephesus 171

Antipholus of Syracuse 170–1, 213, 220, 221
Anti-Racist Shakespeare 189
Antonio 31, 172, 206–7, 215, 216, 217
Antony and Cleopatra 13, 33, 77–8, 125, 126–7, 135, 136
Anusas, Mike 2–3
apparel 168, 169
aristocracy 87, 105, 111, 114, 162, 167, 172, 201, 215
As You Like It 9–10, 19, 53, 104, 105, 119n16, 146–7, 215, 233; of Kenneth Branagh 162–3, 178–80
asexuality 89, 99–100n15
ASMR (Autonomous Sensory Meridian Response) 123
assemblage 161
assembly 12
asymmetrical power relations 5–6, 164
Autolycus 169
Autonomous Sensory Meridian Response (ASMR) 123
Axton, Marie 149, 159n34

Bachelard, Gaston 148–9
Badenoch, Kemi 199, 200

Bailey, Elisabeth Tova 140–2
battle 59, 150, 155
Battle That Gave Birth to the Globe, The 215
Bennett, Susan 29–30
Bentham, Jeremy 2
Best, George 192–3
Bhutto, Fatima 67
Bianca 46, 47, 48, 49, 50
biblical celestial architecture 70–1
Billion Black Anthropocenes or None, A 6
biomorphic transmutation 35
Black History Month 199
Black Lives in the English Archives 1500–1677 187
Black Lives Matter 199
Black Skin, White Masks 195
blazon 95, 221, 222
blood 13, 18, 59–60, 128, 132, 194; and silk 165, 166, 170; and soil 105, 108, 109; and steam 86–7, 87–8, 88–9, 90, 96
Blundeville, Thomas 70–1, 82n21
bodies without surface 77, 79
Book of Skin 193
Borlik, Todd A. 161–2, 163, 173, 181
Bottom, Nick 166–7, 168–9, 174
BP or not BP? 26, 27, 36–7, 39n8
BP plc: logo 28–9, 30, 34, 35, 36–7; RSC sponsorship 26, 27–8, 28–9, 30, 31, 32–3, 33–8
Brackenbury, Sir Robert 131
Branagh, Kenneth 162–3, 178–80
Brathwaite, Peter 186–7, 190, 193–4, 202
Brayton, Dan 124, 125, 127, 130, 131
breath 115, 116, 163, 222; and steam 84, 85, 86, 89, 90, 94, 95, 96, 97, 98
Breen, Philip 212–13, 214, 215, 220

Brennan, Clare 30–1
British Petroleum *see* BP plc

Caliban 37
Callaghan, Dympna 88
canopy the air 72, 73, 74
caterpillars 160, 161, 163, 180–1
Cavert, William M. 54, 56–7
celestial architecture 70–1
celestial globes 70, 71–2, 74, 80, 142, 222
Celia 105, 111, 179
ceremony 58, 59, 109
Cervulle, Maxime 189
chain of being 7, 8, 69
Chapman, George 147
Chaucer, Geoffrey 142, 189
Chiari, Sophie 66–7, 69, 75, 125–6, 126–7
China 29, 172, 177–8
Christian premodern sky 66–7, 68
Christianity 7, 90, 105, 135, 204; and silk 164, 173; and sky 69–70, 71, 76, 80; and snail 147–8, 151–2
Cinthio 176–7
Civil Wars 191
Clash, The 42, 43
climate change 5, 17, 18, 26, 39, 217, 227–8
climate collapse 2, 17, 104
climate consciousness 219
climate emergency activism 122
climate privilege 67
clothing 6, 165, 179
cloud gazing 78, 79
clouds 64, 66, 69, 70, 72–3, 76, 77, 78, 79, 124, 127
Clown 74–5, 80, 81; First 111–13; Second 111
coal 45, 53, 54–5, 56–7
cocoons, of silkworms 160, 163, 166, 174, 180
Cole, Teju 206

Coligny, Admiral 113–14
comedy 45, 57–8, 147, 151, 154, 155, 166; Elizabethan 104, 105, 178; madcap 220; Shakespearean 179
Comedy of Errors, The 32, 118, 220–1, 223; RSC production of 212–13, 213–14, 215; and silk 162, 169, 171, 172, 180
commerce 8, 29–30, 225–6
commercial slime 121, 122
complexion 186–207
Connor, Steven 190, 193
Container Globe, The 226, 227
Conti, Natale 166
contractility, of snails 143
control, politics of 51
Copernicus, Nicolaus 70
Coriolanus 13, 143, 145–6
Countess of Pembroke's Ivychurch, The 91
Covid-19 pandemic 64, 102, 118, 123, 136, 140; and silk 160, 176, 180; and skin 186, 190, 203; and stage 212, 213–14, 220; and steam 84, 85, 86, 98; and surface 10, 11–12, 15, 16–17, 18, 233
cowardice 150, 151
Craik, Katherine 86–7
creation 69–70, 75, 79
creative protest 28, 36, 38
creative writing 142, 175–6, 198
creativity 3–4, 18, 26, 71, 156, 189, 213, 223; and slime 121, 122, 123, 127, 129, 130, 136
Crenshaw, Kimberlé 200
critical medical humanities 2, 4
critical race theory (CRT) 196, 198, 199–200
Critical Studies in Whiteness (CSW) 196
CRT (Critical Race Theory) 196, 198, 199–200

CSW (Critical Studies in Whiteness) 196
Cuckoo's Nest, The 49–50
cultural politics 1, 31, 235
Curtis 43, 46, 55–6
custom 8, 73, 109, 112
Cymbeline 80
Cymbeline Anthropocene 236
Czerski, Helen 18

da Vinci, Leonardo 76–7, 79
Dadabhoy, Ambereen 189
Das Boch vom Schleim 123–4
Davies, Jessica 102
de Acosta, José 125, 126
De Contagione et Contagiosis Morbis et eorum curatione 13
death 35, 45, 67, 191, 199, 224; and silk 169, 173, 174, 175; and slime 125, 130, 131, 133, 136; and soil 106, 108, 109–10, 112, 114, 115; and steam 86, 87, 91, 92, 96; and surface 11, 12, 16, 19
decay 114, 130, 216, 217, 226
decolonisation 189, 197
Deepwater Horizon oil spill 25, 27–8
Denmark, Prince of 72–3, 74, 111, 112
depth 3, 6, 87, 234
Derrida, Jacques 165
Desdemona 135, 174–5, 176, 195–6
digital privilege 12
diseases 11–12, 13, 17, 84, 97, 113, 130, 219
dislimns 77, 78, 83n34
Don Armado 59
Donne, John 221–2, 230n27
Down To Earth 227
dramaturgy 13, 28–9, 38, 50–1, 74, 194, 220, 236; eco- 38, 236

dreams 10, 29, 57, 127, 128–9, 130
Dromio of Syracuse 170–1, 221, 222, 223
Duke of Clarence 125, 130, 131
Duke of Ephesus 169–70
Duke of Gloucester 130–1
Dulac, Anne-Valérie 72–3
Duncan 18, 19
Dyer, Richard 188

Eastward Ho 57
eating 75, 89, 140; see also predation
ecodramaturgy 38, 236
Ecological Approach to Visual Perception, The 64–5
ecological concerns 1, 235
ecological discomfort 219
ecological thought 124
Ecological Thought, The 19–20
economic privilege 16
Eco-Shakespeare and Performance 38
Eklund, Hillary 104
elemental agency 217
Elements of Geometries 4–5
Elizabethan comedy 104, 105, 178
Elizabethan World Picture, The 7
Emmoser, Gerhard 71, 80
environmental activism 26, 39, 104, 236
environmental studies 2, 4, 104
epyllion 89
Estok, Simon C. 117, 122–3, 124
Euclid 4–5
Evelyn, John 54–5
Exterranean 5–6
Exercises, Containing Sixe Treatises 70
extractive industry 32, 122, 173, 177, 181

face masks 11, 85, 160, 233
Fanon, Franz 195
Feng Shui 29

filthy deeds 135, 136
fire 87, 113, 114, 126, 194; and sky 65, 70, 73, 77; and smoke 42, 43, 44, 45, 47, 49, 50, 51, 52, 53–4, 55–6, 59, 60; and stage 222, 229n17; and surface 5, 6, 7, 233, 235
firmament 64–81, 129
First Clown 111–13
fleeting surfaces 52, 57–8
food 12, 25, 34, 46, 102, 104, 117, 122, 153, 227
Forest of Arden 9, 53, 93, 179
fortress city 148
Fossil Free Mischief Festival 36–7
fossil-fuel 25–6, 29, 37, 38, 39, 54, 55, 103–4
Foucault, Michel 2, 233
Fracastoro, Girolamo 13
Fraunce, Abraham 91, 100n19
fuel 43, 49, 53, 54, 60, 86; fossil- 25–6, 29, 37, 38, 39, 54, 55, 103–4
Fulke, William 73–4
Fumifugium 54–5

Gaia principle 7–8
Ganymede 92, 146–7, 147, 179
Garden Theatre 212, 213, 214, 220
geo-humoralism 93, 126, 191
Gibson, James J. 64, 65–6
Gibson, William 3
Gil Hecatommithi 175–7
Glas, Aarie 15–16, 23n38
Gleyzon, François-Xavier 143–4
global selfhoods 126–7
global sky 81
globalisation 223
Globe Ensemble 218–19
Globe Theatre 51, 217–18, 222
Globe-al Dominance 226
Godwin, Simon 117–18

goodly gallery with a most pleasant prospect, into the garden of natural contemplation, a[1] 73–4
Great Chain of Being 7, 8, 69
Great Rebuilding 50–1
Great White Bard, The 177
Gregory, Miles 224–5, 226
Ground-Work 104
Grumio 43, 46, 50, 55–6
Guerra, Debbie 52

Habib, Imtiaz 187
Hakluytus posthumus 142
Hall, Kim F. 188, 192–3, 197
Hall, Stuart 188, 193, 208n7
Hamlet 4, 129, 138n34, 143, 212; and sky 72–3, 74, 75, 76, 77, 79, 80, 81; and soil 105, 111–13, 114, 116, 118, 119n16
heat 29, 50, 67, 118, 126, 174; and steam 86, 87, 92, 94–5, 96
Helios 29, 30
Henry IV Part One 60, 103
Henry IV Part Two 53, 60, 167
Henry V 60, 215, 224
Henry VI Part One 1, 32, 58
Henry VIII 53, 168
Henry VIII: All Is True 51, 58, 217
Herrick, Robert 162, 180
Hidden Killers of the Tudor Home 52
Hollywood Shakespeare 203–4, 204–5, 206–7
Homer 152
Horatio 112
Hortensio 46, 47
Horwell, Claire 85
Hoskins, G.W. 50
human brain 74
human privilege 220–1, 227
human selfhoods 66, 190–1
human vulnerability 86, 223

humanism 5–6, 71, 103, 105, 129, 150, 151, 152, 164
humoralism 13–14, 68, 73, 87, 129, 201, 213, 225; geo- 93, 126, 191; and smoke 44–5, 46, 47; and snail 143, 152–3, 155–6; and soil 105, 114, 116

Iago 135, 175, 195–6
ideological privilege 16
Iliad 152
imagery 9, 10, 25, 27, 28, 30, 32, 72, 78, 110, 127, 134, 142–3, 145, 148, 172, 177, 190, 202
Ingold, Tim 3, 64–5, 65–6, 81, 97, 98
international travel 102
intersectionality 200
Iron Age degradation, of the earth 34

Japan 178–9, 179–80
Jaques 147, 215, 216, 217
John of Gaunt 106, 109
Johnson, Paul 16, 17
Johnston, Mark Albert 150, 151
Jolas, Maria 148–9
Jonson, Ben 57–8
Julius Caesar 60, 72, 86, 87

Kamil, Neil 149
Karim-Cooper, Farah 177, 197
Katherine 44, 45, 46, 47–8, 49, 50, 53, 56
Kaul, Mythili 198
Keats, John 145
Killer Rabbit of Caerbannog 154
King John 59
King Lear 10, 11, 13, 60, 72, 86, 87, 143, 162
Krauze, Andrzej 9–10, 22n23

---

[1] The fulsome title of William Fulke's publication has been edited for indexing purposes.

Lander Johnson, Bonnie 105
landscapes 9, 28, 34, 78, 93, 102, 106, 109, 132–3, 202, 203, 204
Latour, Bruno 8, 9, 161, 227, 228
Lavinia 59, 133, 134
Levinas, Lemnius 191
Lewis, Dyani 85
Lewis, Rhodri 78–9
Lewis, Simon L. 5, 60
light pollution 66
Lipscomb, Suzannah 52
liquefaction 162
Little Jr, Arthur 202
lockdowns 11, 12, 14–15, 64, 123, 140, 214, 218
London's Burning (The Clash) 42
London's Burning (nursery rhyme) 42–3
love 48, 61n10, 145, 146, 152, 216; and silk 161, 174–5, 178; and skin 194–5, 198, 201, 204; and steam 86, 88, 89, 91, 93
Lovelock, James 7–8, 227–8
Love's Labour's Lost 10, 59, 103, 143, 145, 146
Love's Schoolmaster 165
Lucan 191
Lucentio 46, 49, 50
Lucius 59, 133, 134

Macbeth 13, 18–19, 33, 59–60, 86, 87
McCarthy, Harry 189
madcap comedy 220
Magigi, Natasha 202–3
Malvolio 167, 201, 202, 204, 205
man-as-snail 155
Mandel, Emily St. John 11
Manning, Michelle 226
Margulis, Lyn 7–8, 227–8, 234
Marlowe, Christopher 113
Marston, John 57
martial manhood 150

Martin, Randall 38, 53, 91, 104, 215, 228n5, 236
masculinity 145–6, 147, 155
masks: face 11, 85, 160, 233; plague- 10; surgical 213
Maslin, Mark A. 5, 60
massacre 5, 114
Massacre at Paris, The 113, 114
May, Theresa J. 38
May Day 147
mazéd world 219, 220
Measure for Measure 1, 224
medium (Gibson's surface) 65–6
Mehdizadeh, Nedda 189
Mentz, Steve 31–2
Merchant of Venice, The 31, 162, 169, 171, 172, 180, 194, 200, 215
Merry Wives of Windsor, The 49, 53–4
Metamorphoses 6–7, 34–5, 35–6, 86, 90–1, 92, 133, 135, 165–6
metaphor 6, 32, 44, 89, 102, 117–18, 124
metonymy 6, 133
Middleton, Thomas 174
Midsummer Night's Dream, A 14, 162, 167, 168–9, 180, 218–19
Miles 150, 154
Minola, Katherine 44, 45
Mistress Quickly 53, 54
Moffett, Thomas 163–6, 167, 172, 173–4, 175–6, 178, 182n14, 182n18
Monty Python and the Holy Grail 154
Morton, Timothy 19–20
Moth 59
Much Ado About Nothing 202, 203
mucin 141
mud 48, 65, 77, 105, 119n16, 121, 126–7, 135, 219
mulberry trees 161, 165, 172, 173, 177–8
Mulciber 151, 152–3
multiculturalism 202, 203, 205, 207
myth 71, 90, 91, 92, 133, 166

Mythologiae 166
Myxocene 122, 233

narrator, of a poem 90, 95, 96, 144, 162, 222
Nashe, Thomas 128–9, 130, 141
Nell 221, 222, 223, 226
Newton, Thomas 191, 192, 193
nightmares 29, 125, 130, 131–2
nighttime travel 140
Nile, river 125, 127, 128
Nutton, Vivian 97

object studies 2, 4
ocean 17, 25, 28, 32, 67, 102, 122, 125, 199, 203–4
O'Connor, Marion 224
oil imagery 32–3
oil industry 26, 27, 32–3, 37
oil ontology 25–6
Old Shepherd's son 74–5, 80, 81
Olivia 167, 201, 202, 204, 205
O'Malley, Evelyn 37, 38
On Contagion, Contagious Diseases and Treatment 13
On the Revolutions of the Celestial Spheres 70
oppression 199, 200
Orbis hypothesis 5–6
Orientalism 80
Orsino 201, 204, 207
Othello 80, 124, 125, 135–6, 162, 174–7, 180, 194, 195–6, 197–9
Othello: New Essays by Black Writers 198–9
Othello's Black Handkerchief 176
Ovid 6–7, 34–5, 35–6, 86, 90, 92, 165–6

Pakistan flooding 67–8
Palissy, Bernard 148–9
pandemic: Covid-19 *see* Covid-19 pandemic; future 17–18
Panopticon 2

Park, Monomi 122
patriarchy 49, 79, 104
Pauly, Daniel 122
PCRS (Premodern Critical Race Studies) 196, 198, 199–200
perception 3, 46, 65, 74–5, 145, 149, 177, 237
Pericles 31–2
Petruccio 44, 46–7, 47–8, 48, 49–50, 51, 55–6
plague 10, 12–13, 58, 86, 96, 97, 101n31
plants 7, 90, 96, 103, 104, 172, 173, 216; *see also* vegetation
poetics 6, 59, 109
Poetics of Space, The 148–9
pollution 56–7, 58–9, 60, 66, 118, 206, 215
Polonius 76, 77
Pop-up Globes 214, 223–4, 224–5, 226
Portia 194, 195
positionality 15–16, 23n38, 196, 197
poststructuralism 143–4, 188
power 81, 108, 161, 171, 204, 214, 222; over ecologies 72; of elemental agency 217; racial 189, 191, 196, 198, 200; relations of 5–6, 164; and sex 45; of Shakespeare 225; and surfaces 2, 7; textile 162
predation 45, 89–90; *see also* eating
Preedy, Chloe 58–9, 60
Premodern Critical Race Studies (PCRS) 196, 198, 199–200
premodern cultural politics 1, 235
premodern sky 66–7, 68
Prince of Morocco 194–5
privilege 6, 15, 19, 171; climate 67; digital 12; economic 16; human 220–1, 227; ideological 16; politics of 53; social 200;

white 189, 190, 196, 200, 206, 207, 214
props 6, 30, 49–50, 111, 116
Prose of Things, The 4
Prospero 29, 31, 37, 78
Ptolemy, Claudius 71, 79, 80, 126
Purchas, Samuel 142
Purchas his pilgrims 142
Pyne, Stephen J. 233
Pyramus and Thisbe 165, 166
Pyrocene 233

race 6, 164, 176, 221, 222; and skin 188, 191, 193, 196–7, 197–8, 199, 200, 202, 203, 207, 209n26
Race the Floating Signifier 188
RaceB4Race 197–8
racecraft 193, 200, 209n26
racial power 189, 191, 196, 198, 200
racism 189, 194, 196, 200, 203, 207
Rape of Lucrece, The 59, 60
Rebhorn, Wayne A. 89, 90
Recepte Véritable, La 148
reconstructive Shakespeare 224
Rediscovering Black Portraiture 186
Renaissance 117, 130, 143–4, 186–7, 187–8, 189
replicas, of Shakespeare's Globe Theatre 223
respiration 84, 85, 86, 89, 90, 95, 97, 116
Revenger's Tragedy, The 174
rhetoric 4, 59, 74, 106, 110, 196
Richard II 72, 76, 105–6, 109–11, 114, 118
Richard III 125, 128, 130, 131, 134–5, 136
ritual 3, 59, 106, 109
Roman tragedy 126–7, 145–6, 194
Romeo and Juliet 13, 79–80, 198
Rosalind 146–7, 179

Royal Shakespeare Company (RSC): and BP 26, 27–9, 30–1, 32–9; production of *As You Like It* 119n16; production of *The Comedy of Errors* 212, 213–14, 215, 220; production of *Hamlet* 119n16; production of *Much Ado About Nothing* 189–90, 202; production of *Timon of Athens* 114–15, 118; production of *Twelfth Night* 30–1

sacrificing fire 59
Said, Edward 80
Salisbury, Matthew 114
sea images 31–2
Second Clown 111
self-reflection 4, 123, 194, 215
semiotics 124, 136
set design 6
sex 44–5, 86, 89, 97
sexual aggression 144–5
sexual assault 133
sexual desire 95, 147, 201
sexual politics 49, 96, 155
sexual symbolism 144
sexuality 146, 147, 154, 200; *see also* asexuality
sexually charged dialogue 146, 221–2
Shakespeare and Ecology 236
Shakespeare Bulletin 38
Shakespeare Film Company 178, 179
Shakespearean comedy 179
Shakespearean surfaces 1, 3–4, 9, 11, 18, 102, 128, 166, 235
Shakespeare's Gaia 228
Shakespeare's Globe Theatre 49–50, 51, 215, 216, 217, 218, 223, 224, 227–8
Shakespeare's Imagery and What It Tells Us 142–3
Shakespeare's Spiral 143–4

Shand-Baptiste, Kube 200
Shanty Productions 205, 206–7
shape-shifting 19, 35, 77
Shine, Tara 17–18
Shipwreck Trilogy 28, 118
Sidney, Philip 19
signs 6, 29, 47, 59, 68, 76, 77, 134, 136, 146
silk 2, 124, 160–81, 220, 234
Silkewormes and their Flies, The 163, 164, 165, 172
silkworms 161, 163, 164–5, 165–6, 172–4, 175, 176, 181; cocoons of 160, 163, 166, 173–4, 180
simile 6, 143, 144–5, 146, 156
Simonetti, Cristián 2–3
Sir Toby 205
Skewes, Juan Carlos 52
skin 2, 3, 18, 19, 126, 141, 186–207, 234
skin colour 187–8, 194, 195, 196, 203
skin of the living world 118
sky 2, 10, 59, 64–81, 129, 212, 216, 222, 225, 234; Christian premodern 66–7; global 81
slick 2, 25–39, 234
slime 2, 121–36, 144, 164, 234
Slime: A Natural History 123–4
slime dynamics 123–4
slime economy 123–4
Slime Rancher 122
Smith, Ian 176–7
smoke 2, 3, 42–60, 77, 86, 87, 96, 124, 207, 217, 233, 234
smoke pollution 56–7
snail 2, 140–56, 234; contractility of 143
snail imagery 145, 148
snail scrutiny 142
snail secretion 141
snail tentacles 141
snail-as-snail 155

social activism 2, 4
social hierarchy 49, 55, 111
social inequalities 16–17, 31, 122–3
social justice 36, 202, 203
social order 151–2
social privilege 200
social relocation 49–50
social status 74, 75, 220
social surveillance 2
Soedirgo, Jessica 15–16, 23n38
soil 2, 65, 102–18, 121, 124, 234
Soilscape Tragedies 118
space 12, 75, 84, 109, 140, 142, 179, 186; and slime 121, 124, 131; and stage 212, 213, 216, 220, 225, 226
space travel 25
spirit 55, 73, 90, 105, 112, 128, 147, 148, 164, 174–5
Sponsoring Shakespeare 30
sponsorship, of RSC by BP 26, 27, 30, 33, 34, 36–7, 38, 214
Spurgeon, Caroline 31, 142–3, 144
stage 2, 212–28, 234
stage directions 6, 113, 114, 149, 151, 154, 155
stars 66, 69, 70, 71, 72, 76, 79–80, 96, 142, 216
Station Eleven 11
steam 2, 59–60, 84–98, 115, 124, 234
steamy surface 85–6
Stern, Tiffany 216
stockings 167
Storm Desmond 67
Stratford-upon-Avon 12, 13, 16, 30, 35, 36, 38–9, 52–3, 212, 223
structural racism 189, 200, 203
structuralism 65, 223; post-143–4, 188
substance (Gibson's surface) 65–6

sun 54–5, 69, 70, 72, 93, 94–5, 116, 175, 194, 195, 203–4
sun god 29, 30
surface 1–20; bodies without 77, 79; naming of 2; steamy 85–6
surface futures 233–5
surface studies 2–3
Surfaces: A History 3–4
Surfaces: Transformations of Body, Materials and Earth 2–3
surgical masks 213
Symbiocene 233–4
synecdoche 6, 10

Taming of the Shrew, The 42, 43–4, 45, 46–7, 49–50, 52–4, 55–6, 58, 60, 61n5, 61n10
tears 87, 95, 105, 110, 132, 133, 134
Tempest, The 10, 27–30, 32, 37, 78, 118, 214
Terrors of the Night, The 128–9
textilic power 162
textual space 216
Thersites 149–56
Things of Darkness 188, 192–3
Thompson, Ayanna 193, 197–9, 209n26
Tillyard, E.M.W. 7
Timon of Athens 13, 105, 114–15, 115–16, 116–17, 117–18
Titania 218–19
Titus Andronicus 59, 72, 194; and slime 125, 128, 130, 131–2, 132–3, 133–4, 135, 136
tongue 9, 18–19, 133, 134, 135, 144, 157n16
Tosh, Will, 202–3
Touchstone of Complexions, The 191, 192
tragedy 18, 19, 25, 31, 68, 72, 78, 86, 115, 170, 176, 198; Roman 126–7, 145–6, 194
transformative thresholds 3

transience 58, 78
transmutation 58, 91; biomorphic 35
travel 12, 15, 16–17, 31, 54–5, 142, 214, 221; air 64; international 102; nighttime 140; space 25
Troilus and Cressida (Shakespeare) 8–9
Troilus and Crisyede (Chaucer) 142
Türkoğlu, Ayça 123–4
Twelfth Night 28, 30–1, 32, 118, 167, 201, 202–7; multicultural film versions of 203–4; RSC production of 30–1
Twitter (now X) 15, 186
Two Gentlemen of Verona, The 79

Ulysses 8–9
Uncle Marcus 59
Unfortunate Traveller, The 141
Usher, Philip John 5–6

vapours 55, 73, 74, 77, 78, 88, 93, 94, 96, 128, 129
vegetation 10, 88; *see also* plants
ventilation 12, 84, 85
Venus 45, 144, 145, 146; and steam 86, 88–90, 91, 92–3, 94–5, 95–6, 98
Venus and Adonis 32, 44–5, 86, 87–90, 91, 92–8, 143, 144–5
victimhood 199, 200
Vincent, Clare 80
Viola 201, 202–3, 205, 206
viruses 11, 97; corona 11, 12, 16, 84, 85, 160, 180, 213, 233
Visible Skin 186–7, 187–8, 194, 195, 202
Vulcan 151, 152–3

Wall, Cynthia Sundberg 4
water 6–7, 25, 87, 93, 102, 113–14, 160, 180, 207, 214,

235; and sky 65, 67, 77, 78, 79; and slime 121, 125, 128; and smoke 42, 43, 55, 56
weather 31, 64, 66, 67, 68, 75, 91, 94, 110
weather-world 66, 68, 81
Wedlich, Susanne 123–4
White 188
white privilege 189, 190, 196, 200, 206, 207, 214
white saviour industrial complex 206
whiteness 19, 176, 222; of skin 187, 188, 189, 190, 195, 196, 197, 199, 200, 203, 207

Widow 46–7, 48, 50
Wilson, Jeffrey R. 222
Winter's Tale, The 72, 74–5, 162–3, 169, 180
Wohlleben, Peter 102–3
Worthen, W.B. 222–3, 229n17
Wotton, Sir Henry 51–2, 56, 58, 217
Wyckoff, William C. 175

X (formerly Twitter) 15, 186

Yusof, Kathryn 6

Zeffirelli, Franco 50

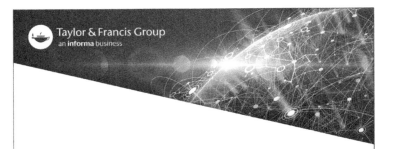